Microcalorimetric and Microbiological *in vitro* Investigations on the Acaricidal, Insecticidal, and Antimicrobial Effects of Propolis

Mikrokalorimetrische und mikrobiologische *in vitro* Untersuchungen auf akarizide, insektizide und antibiotische Wirkungen von Propolis

Inaugural-Dissertation
zur Erlangung des Doktorgrades
am Fachbereich Biologie, Chemie und Pharmazie
der Freien Universität Berlin

vorgelegt von

Assegid Garedew

Berlin 2003

Cover: a propolis collector bee with propolis load on the corbicula

Gedruckt mit Unterstützung des Deutschen Akademischen Austauschdienstes (DAAD)

Die vorliegende Arbeit wurde in der Zeit von April 2000 bis August 2003 an der Freien Universität Berlin, Fachbereich Biologie, Chemie, und Pharmazie, am Institut für Biologie/Zoologie unter der Anleitung von Prof. Dr. Ingolf Lamprecht, Dr. Erik Schmolz und Prof. Dr. Burkhard Schricker angefertigt.

Bibliografische Information Der Deutschen Bibliothek

Die Deutsche Bibliothek verzeichnet diese Publikation in der Deutschen Nationalbibliografie; detaillierte bibliografische Daten sind im Internet über http://dnb.ddb.de abrufbar.

ISBN 3-8325-0431-1

Logos Verlag Berlin
Comeniushof, Gubener Str. 47,
10243 Berlin
Tel.: +49 030 42 85 10 90
Fax: +49 030 42 85 10 92
INTERNET: http://www.logos-verlag.de

1. Gutachter: Prof. Dr. I. Lamprecht
2. Gutachter: Prof. Dr. K. Hausmann / Prof. Dr. B. Schricker

Datum der Disputation 13.11.2003

To my son Lucas

It must be realized that empiricism has its limits, and we must face the fact that a deeper understanding of biology may require a more heuristic approach. The making of new theories is, however, a risky business and much stumbling can be anticipated along the way to successful, general, and fruitful results.

H.J. Morowitz (1968)

Table of Contents

1.0 General Introduction

In this thesis three main subjects are integrated: the impact on honeybees and the possible means of control of the honeybee parasites and pests *Varroa destructor* (Anderson and Trueman), and *Galleria mellonella* L.; the natural bee product propolis, and its biocidal use; calorimetry as a technique applied in the investigations of metabolic rates and the sublethal effects of propolis. As the thesis is an interdisciplinary approach involving the mentioned fields, it is important to give concise introductions to the main subjects. The intention of these introductions is that a calorimetrist reading this thesis can gain some general ideas about the other subjects in the area of bee research. The same holds true for a bee researcher to whom, in most cases, the what about and the working principles of calorimeters are not familiar, reading this thesis without a proper introduction of calorimetry may make it difficult to grasp what is being conveyed.

1.1 Varroa mites as parasites of honeybees

The hive of honeybees with its constantly maintained optimal temperature, humidity, and carbon dioxide level, year round ample availability of the host bees, protinacious (pollen), carbohydrate (honey), and wax foods, is a suitable habitat for a diverse array of parasites and pathogens (Bailey and Ball 1991). Some of the most common parasites and pathogens of the honeybees include viruses (acute paralysis virus – APV, deformed wing virus – DWV, and sack brood virus - SBV), bacteria (*Paenibacillus larvae larvae* - American foulbrood, and *Melissococcus pluton* – European foulbrood), fungi (*Ascosphaera apis* – chalkbrood, and *Aspergillus flavus* – stonebrood), protists (*Nosema apis* - nosema disease, and *Malphigamoeba mellifica* - amocba disease), mites (*Varroa destructor*- varroosis, and *Acarapis woodi* – tracheal mites), and insects (*Galleria mellonella* – the greater wax moth, and *Aethina tumida* - the small hive beetle). Among the different parasites and pathogens mentioned, the parasitic mite *Varroa destructor* (Anderson and Trueman, formerly called *Varroa jacobsoni* Oud.) is becoming a global concern affecting the beekeeping industry based on *Apis mellifera* L. (Boecking and Spivak 1999), and it is attracting the attention of researchers to circumvent the perish of the honcybee. It is not only the beekeeping industry that suffers from loss of the honeybees; rather the crop agricultural sector is also being hit by this problem, because most plants are dependent on bees for pollination. It is estimated that 80% of all crop insect pollinations are accomplished by honeybees (Benedek 1985).

The infestation of *Apis mellifera* L. by *Varroa destructor* reportedly originated nearly half a century ago (Smirnov 1978, Crane 1979, Matheson 1995), when the mites transferred to *A.*

mellifera colonies that had been introduced into the home range of *A. cerana* Fab., the mite's original host. *Varroa destructor* is an obligate ectoparasite that feeds on the hemolymph of bees both in the capped developmental stage and on adults, but reproduces only in the capped worker and drone brood of *A. mellifera,* and only in the drone brood of *Apis cerana.* In its original host, the Asian honeybee *Apis cerana*, a host-parasite relationship has evolved that rarely damages the host (Anderson and Trueman 2000). In the case of *A. mellifera* colonies, however, mortality from *V. destructor* can reach up to 100% within two to five years, if mite control methods are not implemented (De Jong 1997). Additionally, high mite populations were observed to be associated with increased incidences of viral infections (Ball 1994), lower weight at hatching, and shortened life span of the adult bees (De Jong et al. 1982), as well as deformed wing and shortened abdomen.

The vertical transmission of Varroa mites from individuals of the parent to those of the offspring colony involves the formation of a daughter colony with parasites from the parent colony after swarming, or the splitting of parent colony by the beekeeper.

The extent of the problem of varroosis is alarming mainly due to the very high spread potential and debilitating action of the parasitic mites. The very close contact between bees in a colony facilitates the easy intracolony spread of the parasite among individuals within a generation (horizontal transmission). This adds up to the likely demise of the colony, should even a single member of it is infested. The very high horizontal intercolony dispersion potential of *V. destructor* can be attributed to at least two main factors: firstly, because of activities of beekeepers moving colonies from place to place for commercial and pollination purposes; secondly, due to intercolony drifting and robbery of infested bees. The intercolony drifting of infested bees and the spread of varroosis is worsened by the repercussion effect of the weakening of colony-state factors, and, thus, behavioural change of the individual workers by the parasitic mite (Downey et al. 2000). Several stress events, such as wax deprivation (due to its insufficient production as a result of underdeveloped glands of the infested bees), depletion of nectar and pollen stores, and worker loss (Winston and Fergusson 1985) induce resource gathering responses in honeybee colonies. Such responses include an increased number of foragers, accelerated task ontogeny (i.e., earlier onset of foraging flights), or a greater effort by individual foragers (visiting more flowers or carrying larger pollen, nectar, or propolis loads) (Schmid-Hempel et al. 1993). Among the different responses of the colony to the stress imposed on it, the precocious foraging (accelerated task ontogeny) contributes to the increased horizontal transmission of parasites from an infested colony to another one by the increased drifting of parasitized and weakened workers. It was confirmed by Schneider and Drescher (1988) that worker bees parasitized by Varroa mites during their pupal development start flying earlier, and the rate of drifting of such bees was found to be very high compared to drifting by non-infested bees. Bowen-Walker and Gunn (2001) explained the possible reason for the higher drifting rate in infested colonies to be due to the fact that by flying earlier in their lives, infested bees start out nest activities before their memory/orientation is fully developed, leading to their disorientation and loss.

1.1.1 Biology of *Varroa destructor* mites

The female *V. destructor* mite is brown to reddish-brown in colour, measuring 1.1 to 1.2 mm in length and 1.5 to 1.6 mm in width (about the size of a pinhead) (Fig. 1.1 a and b). Males are smaller, about 0.7 mm by 0.7 mm, and light tan in colour. Even though the female mite parasitizes larval, pupal, and adult developmental stages of the honeybee, reproduction takes place only in the capped brood developmental stage (Infantidis 1983, Martin 1994, Steiner et al. 1994). This reproduction lasts 12 days in worker and 15 days in drone brood (Moritz 1985, Le Conte and Cornuet 1989). Outside the capped brood, the female Varroa mites live on adults, mostly on nurse bees, using them mainly as short term hosts and for dispersal (phoresy); for this reason mites on adult bees are called phoretic mites. It was forwarded by Hoppe and Ritter (1989) that Varroa mites prefer young "house" bees to older worker bees, probably due to the lower titer of the Nasonov gland pheromone geraniol, which strongly repels the mite. When on adult bees the mite fits itself beneath the bee's abdominal sclerites, lessening transpirational water loss, and reducing the vulnerability to grooming and dislodgement during host activity (Sammataro et al. 2000). The dorsoventrally compressed body shape of the mite allows it to fit properly into the intersegmental groove, at the same time accessing the soft integument that can be pierced by the mite's chelicerae, enabling it to feed on the bee's hemolymph. Males are not able to pierce even the soft integument of the brood stage, since their mouth part is modified for sperm transfer (Frazier 2000). As a result, male mites are dependent on the hole made by the female mites to suck hemolymph from the brood stage. In addition to that, the body structure of male mites is not optimally compressed to fit under the abdominal sclerites of the adults. For these reasons the lifespan of male mites is restricted only to the capped brood developmental stage.

One or sometimes more foundress Varroa mites enter the prepupal stage of a worker brood 20 h, and a drone brood 45 h before cell capping in order to reproduce (Boot et al. 1991), and start feeding on the brood and its reserve food. The time at which the mother mites enter the uncapped brood indicates the period of attractiveness of the brood to mites, since only brood of a particular age is attractive to them (Fries et al. 1994). Drone brood was found out to be more attractive to Varroa mites than worker brood (Fuchs 1990). The higher preference of drone brood to worker brood by Varroa mites was considered to be the result of a combination of several factors. These factors include chemical attractants, such as fatty acid esters secreted by the larvae, and present on the cuticle of drone brood at higher quantity and/or quality than in worker brood, at the mite attractive age of the brood (Le Conte et al. 1989, Trouiller et al. 1991, Beetsma et al. 1999, Sammataro et al. 2000). Non-chemical factors include longer period of pre-capped Varroa attractive stage (Fuchs and Müller 1988, Infantidis 1988, Boot et al. 1991) as well as bigger size of the drone brood cell, which increases the chance of encounter (Sammataro et al. 2000). In addition to that, broods that are big and grown up to the rim of the brood cell are more attractive than broods far from the rim (Beetsma et al. 1999).

After entering uncapped brood cells the mites hide from the removal action of the nurse bees by submerging into the liquid brood food until cell capping. While in the submerged state, they use their peritremes (Fig. 1.1 c, and d) which protrude snorkel-like above the liquid food for respiration (Donzé and Guerin 1997). Mites emerge out of their concealment after brood capping. A foundress mite lays its first egg, which develops to a haploid male (n = 7), 60 h after cell capping, and the subsequent eggs are laid in intervals of 30 h and develop to diploid females (2n = 14) (Steiner et al. 1982, De Ruijter and Pappas 1983, Infantidis 1983, Rehm and Ritter 1989).

The male mite requires 6.9 days to reach the adult stage whereas a female needs only 6.2 days (Rehm and Ritter 1989). The male mite is already a mature adult by the time its sisters reach maturity, and it copulates with all of its adult sisters as often as possible, to ensure fertilization, before the bee emerges as a callow bee (Sammataro 2000). After the bee has hatched and left the cell with the mature female mites, the male and nymph stages of the female mites die of starvation. The number of female mites that reach maturity is directly affected, among other factors, by the length of post capping developmental period. Since the drone brood has a longer post capping developmental period of 15 days, compared to the 12 days for workers (Moritz 1985, Le Conte and Cornuet 1989) higher numbers of female mites emerge as adults from drone than from worker cells.

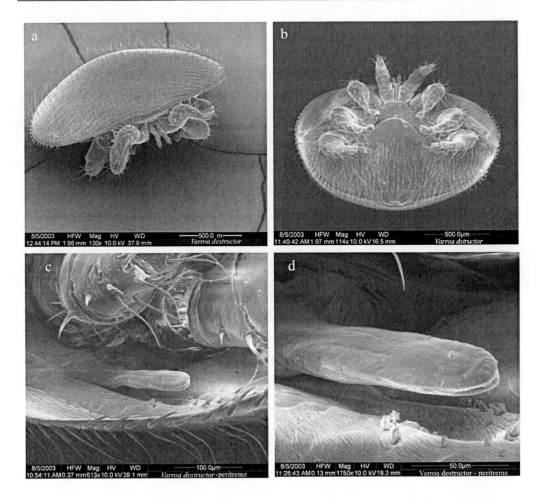

Fig. 1.1 Scanning electron microscopic pictures of the female *Varroa destructor,* (a) dorso-
frontal view, (b) ventral view, (c) the breathing structures stigma and pertitreme, and
(d) an enlarged peritreme with the slit. (FEI, Quanta 200).

1.1.2 Defence mechanisms of *A. mellifera*

In general, the defence mechanisms of honeybees to protect themselves from pathogens
and parasites include (a) a constitutional defence - the chitinous exoskeleton, (b) cellular and
humoral defences - haemocytes, and enzymes and antimicrobial factors, respectively, (c) a
physiological defence –the proventricular valve that filters ingested spores, and (d) a behavioural
defence - activity of the honeybees to keep themselves, their nest mates, and their hive clean; an
important mechanism to stop the spread of pathogens such as American foulbrood, chalk brood,
and parasites. The behavioural defence mechanism is of special interest from the perspective of
the bee's defence mechanism against parasites such as *Varroa destructor.*

1.1.3 Behavioural defence of *Apis mellifera* against *Varroa destructor*

The behavioural defence mechanisms of *Apis mellifera* that enable it to reduce the population of *V. destructor* mites are hygienic and grooming behaviours (Boecking and Spivak 1999). The expression of these two behavioural traits in *A. mellifera* is far lower compared to that in *Apis cerana,* the original host of Varroa mites (Peng et al. 1987, Boecking and Ritter 1993, Shimanuki et al. 1994). The higher grooming and hygienic activity of *A. cerana* combined with the non-fertility of mites in worker brood endowed tolerance to this bee species enabling it to live in equilibrium with Varroa mites (Fries et al. 1994, Boecking and Spivak 1999).

Regardless of the effort of a colony to keep its hive clean and free of parasites and pests, which actually varies from race to race (Buchler 1994) and is influenced by climatic conditions (De Jong et al. 1984, Kraus and Velthuis 1997), beekeeping with the western honeybee *A. mellifera* is being highly jeopardized by *V. destructor*. Though some fragmentary and anecdotal reports exist about the tolerance of some *A. mellifera* colonies to *V. destructor* from different regions, the only race of the western honeybee which is confirmed to be tolerant to varroosis and does not need human interference is the Africanized honeybee (De Jong et al. 1984, De Jong 1996, Medina and Martin 1999). Factors that are involved in the resistance of varroosis by Africanized honeybees include reduced mite fertility (Martin et al. 1997), higher offspring mortality (Medina and Martin 1999), smaller brood cell and hence limiting space (Message and Gonçalves 1995), shorter post capping developmental period (Moritz 1985), and food (pollen) availability (Moretto et al. 1997). In addition to these, behavioural factors of the bees such as hygienic (Corrêa-Marques and De Jong 1998) and grooming (Moretto et al. 1993) behaviours were deemed crucial.

The threat of *V. destructor* to the non-Africanized western honeybees is so alarming that colonies have to be somehow treated or manipulated in order to reduce the population size of Varroa mites and to save the colony from dying out.

1.1.4 Control methods of *Varroa destructor*

Different methods of treatment of a colony are available nowadays, even though some of them are ineffective and others have limitations due to their effects on the bees or the beekeeper. The methods of Varroa prevention and control include biotechnical, biological, and chemical methods.

1.1.4.1 Biological methods

The biological Varroa control methods involve the use of the bee's biology, perhaps its natural resistance against mites. The desirable features of bees that can be selected to establish a resistant colony include higher hygienic and grooming activities, shorter post capping periods, low attractiveness of brood to mites, and low mite fecundity factors. The selection and establishment of resistant colonies is the best and cheapest method of control of varroosis since the bees themselves deal with Varroa mites. Achievement of this control method is, however, taking longer time and short term solutions, such as biotechnical or chemical methods have to be used in the meantime to stop colony death.

1.1.4.2 Biotechnical methods

Biotechnical methods of mite control utilize the principle that mites inside a capped brood are trapped and hence can be removed from the colony. The drone brood, which is often unwanted by the beekeeper, can be used as a trap comb. In the period of absence or scarcity of drone brood, worker brood can also be used as a trap. It is, however, undesirable to destroy worker brood with the trapped mites; the mites have to be killed selectively. The selective killing of mites can be done at a high temperature (44 °C) (Rosenkranz 1987, Engels 1994), and with the use of formic acid (Fries 1991, Calis et al 1998)

1.1.4.3 Chemical methods

The chemical methods of mite control involve various methods of application and ways of dispersal of the acaricides, which are determined by the nature of the chemicals being used. The methods of application include: (a) hanging impregnated plastic strips between combs in the brood chamber. The chemicals are distributed among members of the colony by contact of some bees with the impregnated strip and subsequently with their nest mates. The crowded life style and close contact among bees is responsible for the distribution of acaricides applied this way. This method is used to apply Bayvarol™ (flumethrin as the active ingredient - a synthetic pyrethroid), Apistan™ (fluvalinate as the active ingredient - a synthetic pyrethroid), and Apivar™ (amitraz as the active ingredient). (b) Emulsion or solution in water trickled into bee spaces between combs. Perizin™ emulsion (with coumaphos, an organophosphate, as the active ingredient) and Apitol™ solution (with cymiazol hydrochloride as an active ingredient) are applied this way, and distributed among individuals in a colony by grooming and trophallaxis (food exchange between bees). Organic acids such as oxalic and lactic acids are also trickled on bees in sugar syrup. (c) Feeding acaricides to bees with sugar solutions so that it is distributed by trophallaxis. This method is used in the application of Galecron™/K79 (with chlordimeform as the active ingredient). (d) Smouldering of impregnated cardboard strips in a sealed hive. This method is used to apply Folbex VA™ (with bromopropylate as the active ingredient). Spreading of the active ingredient in the beehive is achieved by combustion. (e) Placing impregnated cardboard on top of the brood combs. Essential oils such as thymol, neem oil, and others are applied this way under the trade name Api-Life VAR™. Evaporation is the means by which distribution is achieved in the beehive. The mechanism by which essential oils act was supposed to be lethality at higher concentrations, and interference with the olfactory senses and orientation of mites at sublethal concentrations (Kraus et al. 1994). Since higher concentrations may also affect honeybees, it may be desirable to use sublethal doses of essential oils. These lower concentrations interfere with the chemoreceptors of the mites, making them unable to locate brood cells to invade and reproduce (Kraus et al. 1994), thus, falling to the bottom and dying of hunger. (f) Evaporation of a solution. Nowadays formic acid is applied in a colony by placing it in an evaporator, which allows the gradual evaporation of the acid.

Though some of the chemicals used for the control of Varroa mites in different parts of the world are toxic to the honeybees and humans, the chemical method of treatment is the only effective and non laborious method presently available to the beekeeper. Several researchers are focusing on the potential use of natural products, such as essential oils for mite control. Even if propolis occurs in the beehive and may not be considered as a contaminant, if used as an acaricide, investigations on its potential use against Varroa mites are lacking. This is despite the fact that propolis showed biocidal activities against a range of microbes, parasites and ailments. One of the aims of this work is, therefore, to investigate the varroacidal actions of propolis.

1.2 *Galleria mellonella* as pest of the honeybees

The wax moth belongs to the subfamily Galleriinae of the family Pyralidae in which the females characteristically lay their eggs in beehives. This subfamily consists of two species known to be pests of the beehive, the greater wax moth *Galleria mellonella* and the lesser wax moth *Achroia grisella*. Both of these species have the same type of scavenging habits, but the lesser wax moth does not cause much damage, and hence is not a serious problem of beekeeping (Charrière and Imdorf 1997). Attention will, therefore, be given to the greater wax moth.

Galleria mellonella can be useful for the beekeeper in some aspects since it could recycle combs of colonies that die in the wild as well as the beeswax combs of the beekeeper. This moth can be reared purposefully as fish bait, animal feed, for scientific research, and it is a model system in insect physiology (Caron 1992). Regardless of its desirable uses in different fields, the wax moth is seen as a honeybee pest by the beekeeper. Normally, the wax moth attacks only abandoned beehives, or active ones in which the bee colony has been weakened, e.g., as a result of disease or starvation. The beekeeper is more likely to see the adult moth, but it is the larval or caterpillar (worm) stage that causes damage to wax comb (Fig. 1.2 a and b). Wax moths fly mainly at night and rest in dark places during day time. They have acute sensory capability to find and exploit beeswax. Wax moths do damage during their larval stages, destroying combs and honey, but adults do not feed since they possess atrophied mouth parts (Charrière and Imdorf 1997). A female starts laying eggs 4 to 10 days after hatching (Shimanuki 1981) and produces 300 - 600 eggs in her lifetime, usually laid in batches of 50 – 150. These eggs are laid in cracks between hive parts, in dark and hidden places (Morse 1978). Wax moth eggs hatch to the larval stage in 5 to 8 days and the newly hatched larvae tunnel into the combs, leaving a complex of silken galleries behind. The larvae chew their way down to the midrib of the comb in order to be safe from patrolling adult honeybees, an important strategy of adaptation for their successful invasion of hives (Caron 1992). The tunnelling destroys the wax cells of the comb and causes leakage of honey due to puncturing of honey storage cells, making honey unmarketable. The larval stage feeds continuously on cocoon, faecal matter of the bee brood, debris, pollen, and wax (though indigestible), and it doubles its weight every day, under ideal conditions, for the first 10 days (Morse 1978). Wax moths prefer impurities in beeswax and, therefore, a comb used for brood rearing is at great risk. The larval and pupal development of a wax moth is aborted if it infests a colony containing only new and foundation combs, or combs used for honey (Morse 1978, Caron 1992, Charrière and Imdorf 1997).

Unlike most other parasites and pests of honeybees, the wax moth causes damage not only in a colony; it also causes destruction of stored combs. Dark and old combs used for brood rearing are the most difficult to store safely since they are full of cocoons and debris, and, hence, ideal to be infested by wax moths. Combs which are new or those used only for honey and stored in dry places have very little appeal to wax moth (Moosbeckhofer 1993). Wax moth larvae are most destructive to beeswax combs in storage, especially in areas that are dark, warm and poorly ventilated (Morse 1978).

In addition to the direct destructive impact caused by the larval stage, adults and larvae could also play roles in transmitting viral, bacterial, and fungal infections from infested to healthy individuals, facilitating demise of a diseased colony (Borchert 1966). Therefore, the control of *Galleria mellonella* in weak colonies and in honeycomb storehouses is very important. Different methods of control are available nowadays, but most of them are accompanied with one or more drawbacks, as will be demonstrated latter in chapter 6.

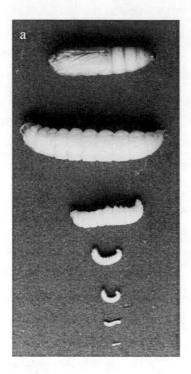 

Fig. 1.2 Larvae and pupa (a), and a male (top) and a female (bottom) adults (b) of the greater
wax moth *Galleria mellonella*

1.3 Propolis

Propolis (bee glue) is a resinous sticky gum collected by honeybees from various plants.
Many plants have evolved mechanisms of protecting their leaves, flowers, fruits, buds, pollen,
and prevent infection of wounds by producing a resinous substance with potent antimicrobial,
anti-putrefaction, waterproof and heat-insulating properties (Münstedt and Zygmunt 2001).
Resin oozes out after injury of a plant part in order to stop further sap loss and prevent infection
of the wound, or it could be actively secreted as a protective covering of buds, to inhibit
sprouting and subsequent death while frost (Crane 1990). Honeybees make use of the result of
this long time evolution of plant secondary metabolism to protect their hives from infection
(König and Dustmann 1988).

Honeybees collect resin from cracks in the bark of trees, leafs, boughs, and leaf buds, and
masticate it by adding salivary enzymes. The masticated gum is then mixed with beeswax and/or
other foreign materials, based on need for further use (Ghisalberti 1979, Marcucci 1995). It has
been noted by Marcucci (1995) that the compounds in propolis originate from three sources:

plant resins collected by bees, secreted substances from bee metabolism (wax and salivary enzymes), and foreign materials which are introduced during propolis elaboration. The relative composition of the three different components varies based on the geographical location of the hive, vegetation composition, bee species, and availability of propolis source plants (Meyer 1956, Johnson et al. 1994, Burdock 1998). The more the available plant resin the less the proportion of wax and foreign materials added to make propolis, and vice versa. In seasons and locations where propolis source plants are scarce colonies suffer from propolis shortage and bees were observed collecting "propolis substituents", like asphalt, paint, and mineral oils (König 1985). These propolis substituents are then mixed with the available resin and used in the beehive. Even under favourable conditions of propolis collection, where there is no shortage of plant resin, the relative proportion of wax added to make propolis is dependent on the purpose for which it is to be used (Meyer 1956, Johnson et al. 1994). Propolis used to repair honey combs is often supplemented with large quantities of wax to give a firmer composition, whilst propolis applied as a thin layer on the internal wall of the hive contains very little or no wax, since wax has no antimicrobial effects (Ghisalberti 1979). The colour of propolis varies from yellow green to dark brown depending on the source plant species and its age. Its consistency is highly affected by temperature: being sticky and pliable above 30 °C, hard and unbreakable at about 15 °C, and brittle and easily pulverizable at a temperature less than 5 °C, especially when frozen (Hausen et al. 1987). The melting point of propolis lies on average between 60 and 70 °C, and it could even go up above 100 °C (Neunaber 1995).

1.3.1 Botanical origin

The fact that bees collect plant resins to prepare propolis was confirmed for the first time by Rösch (1927). Though the botanical origin of propolis was then generally accepted, it was not clear which plant species were used as sources. The difficulty in the identification of the plant species used as propolis sources lay mainly in the fact that propolis collection is a rare event carried out by few bees specialized for this purpose, and also that it often takes place high up in the trees (Crane 1990). The identification of the source plants of propolis involved observation of plants which the propolis collectors visit and comparative chemical analysis of propolis and plant resins (Bankova et al. 1992). The comparative chemical analysis of propolis and resin exudates of trees suspected to be propolis sources (mainly poplar and birch) started at the beginning of the 1970's, and similar chemical compositions between propolis and the corresponding resins were confirmed (Lavie 1976, Popravko 1978).

Nowadays it is commonly accepted and chemically demonstrated that the bud exudates of *Populus spp.* and their hybrids are the main sources of bee glue in temperate zones, such as in Europe (Tamas et al. 1979, Popravko and Sokolov 1980, Nagy et al. 1986, Greenaway et al. 1987), North America (Garcia-Viguera et al. 1993), the non tropical regions of Asia (Bankova et al. 1992), and New Zealand (Markham et al. 1996). In Russia, especially in the northern part, however, the main source of propolis is the birch *Betula verrucosa* (Popravko and Sokolov 1980). Apart from poplar and birch, other plant species, such as conifers (*Pinus spp.*), horse chestnuts *(Aesculus hippocastanu)*, *Prunus spp.* (almond, apricot, cherry, nectarine, peach, and plum trees), willow (*Salix spp.*), alder (*Alnus spp.*), oak (*Quercus spp.*), and hazel trees (*Corylus spp.*) play minor roles as propolis sources in temperate regions (Ghisalberti 1979). In tropical regions there are no poplars and birches and bees have to find propolis source plants (Bankova et al. 2000). The propolis source plants in tropical regions are highly variable due to the immense biodiversity of the flora. For this reason different plant species have been confirmed as propolis sources in various tropical countries. Some of the propolis source plant species, confirmed by observation of the flight activities of propolis collectors and by comparative phytochemical analysis include: *Cistus spp.* (Martos et al. 1997); *Ambrosia deltoidea* (Wollenweber and Buchmann 1997); *Clusia major* and *Clusia minor* (Guttiferae) (Tomas-Barberan et al. 1993); *Acacia spp.*, *Eucalyptus spp.*, and *Xanthorrhoea spp.* (Ghisalberti et al. 1978); *Araucaria spp.* (Bankova et al. 1996, Bankova et al. 1999), *Baccharis spp.* (Banskota et al. 1998, Marcucci et al. 1998, Bankova et al. 1999). Most of the data about the propolis source plants in the tropics relate to Australia, Brazil, and some other South American countries. The plant origin of propolis in most African, tropical Asian, and some South American countries is not yet known.

1.3.2 Collection

Propolis is collected by worker bees that are older than 15 days and specialized for this purpose (Bogdanov 1999). These bees are usually older than those that build comb and cap honey cells, and their wax glands are atrophied (Ghisalberti 1979). Since propolis is hard and difficult to handle at lower temperatures, bees usually collect it in the late afternoon of warm seasons of the year when it is relatively flexible, though very sticky. Bees were observed collecting propolis starting from spring up to early autumn in warmer regions of Europe such as Italy. In most other parts of Europe and the temperate zone in general, the high time for propolis collection was confirmed to be late summer and early autumn, and propolis collection is considered to be a preparation for overwintering (Bogdanov 1999). Due to the sticky nature of propolis, it is not a simple task for the bees to collect it, but the further processing and use in the beehive becomes relatively simpler due to the addition and mixing with salivary gland secretions and wax (Droege 1989). A propolis collector bee may collect and carry up to 10 mg propolis (Fig. 1.3 a and b). Depending on the bee race and the geographical location of the hive, a colony in Europe is able to collect 50 to 150 g propolis per year; the Caucasian bees, however, can collect up to 1000 g (Bogdanov 1999). It is possible, however, to provoke the bees to go for more propolis collection. At present, one of the best methods used for commercial production of propolis is to place a plastic mat with mesh size not more than 4 mm under the top cover inside the hive. Other methods of provoking the bees for more propolisation include sending a drought through a hole in the hive (Bogdanov 1999), placing a mouse dummy, and sending a strong electromagnetic field over the beehive (Horn 1981).

Fig 1.3 A propolis collector bee with propolis load on the corbicula (a), and while "stealing" propolis from a chunk of it left on the hive by the beekeeper (b).

1.3.3 Uses in the beehive

Bees make use of the two important features of propolis in the beehive: mechanical and biological. The mechanical uses of propolis include its application as a thin layer on the inner wall of the hive or other cavities they inhabit. This may prevent loss of moisture in dry seasons (Baier 1969, Möbus 1972) and its catastrophic influx following heavy rainy seasons (Münstedt and Zygmunt 2001), enabling the bee community to keep the hive interior at a desirable moisture level. The presence of propolis as a thin layer also acts as a varnish, smoothing out the internal wall, making it more slippery, and enabling the honeybees to blow off invading ants (Münstedt and Zygmunt 2001). Propolis is also used to block holes and cracks less than 5 mm in diameter (crevices of diameter more than this could be filled-up with wax) (Droege 1989), to repair combs, to strengthen the thin borders of the comb, and for making the entrance of the hive weathertight or easier to defend. This latter mechanical use of propolis might have led to the origin of its name from two words in ancient Greece: *pro* (for, in front, in defence) and *polis* (city, community), to refer to the substance for or in defence of the hive, analogous to walls or fences built around towns/cities to protect them from enemy attack in ancient times. The cape honeybee *Apis mellifera capensis* has been observed using propolis for encapsulation (imprisoning) of the parasitic small hive beetle (SHB) *Aethina tumida*, which could not be killed because of its hard exoskeleton and defensive behaviour, trapping and starving it to death (Neumann et al. 2001). In addition to the mechanical use, the presence of propolis in the beehive also has biological roles; it is used to embalm dead intruders which the bees have killed but could not transport out of the hive, thereby containing putrefaction. It is therefore responsible for the lower incidence of bacteria and moulds within the hive than in the atmosphere outside (Ghisalberti 1979). Propolis is applied as a thin layer on the inner wall of the comb cells before the queen lays eggs, probably to protect the brood from microbial infection (Droege 1989). The presence of propolis at the hive entrance plays not only a mechanical role, but also a biological one in that it acts as a repellent or simply reduces the attention of potential intruders, perhaps disguising the hive chemically as a part of an uninteresting plant (Münstedt and Zygmunt 2001). Propolis also acts as an inhibitor of seed germination and bud sprouting in the beehive, thereby preventing invasion of the hive by plant life (Ghisalberti 1979).

1.3.4 Chemical composition

The chemical make-up of propolis is mainly determined by the resin exudates of plants; the metabolic products of bees i.e. salivary enzymes and wax added to it, as well as foreign materials incorporated during refining play minor roles. Most plant resin components are incorporated into propolis without alterations, but it is likely that some of the components are subject to enzymatic changes by the bees' saliva during the collection or addition of the exudates to bees' wax to make propolis (Greenaway et al. 1990, Burdock 1998). The enzymatic changes may include chopping the carbohydrate components of flavonoid glycones with glucose oxidase to convert them to flavonoid aglycones (Greenaway et al. 1987). The specific chemical composition of propolis is highly influenced by the geographic location of the collection site and the collecting bee species. The largest group of compounds reported in propolis are flavonoid pigments which are ubiquitous in the plant kingdom. There are in general more than 200 hitherto identified compounds that belong to: amino acids, aliphatic acids and their esters, aromatic acids and their esters, alcohols, aldehydes, chalcones, dehydrochalcones, flavanones, flavones, hydrocarbons, ketones, terpenoids and other compounds (Marcucci 1995, Bankova et al. 2000).

Of these compounds, the flavonoids have been the most investigated and were shown to be responsible for the different biological activities ascribed to propolis.

Regardless of the high variation in the specific chemical make-up of propolis collected from different geographic locations and by various bee species or subspecies, its general chemical make-up, under favourable propolis collecting conditions, remains almost the same. It is generally composed of about 50% resin and vegetable balsam (components extractable in ethanol), 30% wax, 10% essential oils, 5% pollen and 5% various other substances including organic debris (Ghislaberti 1979).

1.4 Calorimetry in biological investigations

All physical, physicochemical, chemical and biochemical reactions are associated with the production or consumption of heat and, therefore, with the flow of heat between the system and its surrounding. Calorimeters are instruments used to measure such heat and heat flow rates. Thermodynamically defined, processes that liberate energy such as catabolic cellular reactions are called exothermic/exergonic, and those that absorb energy, such as anabolic/biosynthetic cellular reactions are called endothermic/endergonic. During the catabolic degradation of a substrate into its intermediate or end products, part of the liberated enthalpy is conserved in the production of ATP, and the rest is evolved as heat (Q_{cat}). Part of the energy stored in ATP, which is formed during catabolic reactions, is consumed by biosynthetic (anabolic) reactions, and the rest is given off as heat (Q_{anab}). The net heat production of life process/metabolism (Q_{met}) that one can measure calorimetrically is, thus, the sum of Q_{cat} and Q_{anab}. The calorimetric monitoring of this heat flow between a system and its surrounding could involve analytical calorimetry, whereby the qualitative question whether heat is produced/absorbed or not is answered, or quantitative calorimetry, that measures the amount of heat released/absorbed (Lamprecht et al. 1991).

1.4.1 Types of calorimeters

Different types of calorimeters are in use in various fields of science nowadays, with the classification being done by the use of a combination of several criteria/working principles of the calorimeters. The reader interested in the criteria used, and classification of calorimeters into groups is referred to Hemminger and Sarge (1998).

Calorimeters are commonly divided into batch and flow instruments based on the position of the reaction vessel/fermenter (Lamprecht 1983). Batch calorimeters are those types in which a closed vessel within the calorimeter contains all necessary ingredients for the reaction and perhaps auxiliary equipments for initiating the reaction, stirring, mixing, illuminating etc. In flow calorimeters, however, the reaction occurs in a separate fermenter or vessel, outside the calorimeter, and only a part of the solution is pumped through the flow spiral of the calorimeter, where the heat production rate is measured. The calorimetric experiments with insects and mites in this thesis work were carried out with batch microcalorimeters, whereas all microbiological experiments were done with a flow microcalorimeter.

All the batch and flow microcalorimeters used in this thesis work are based on the heat exchange/conduction principle, whereby the heat produced in the calorimetric vessel is conducted to the surrounding heat sink (isothermal jacket) with an enormous heat capacity, to maintain the temperature of the calorimetric vessel constant. Such calorimeters are still referred by most calorimetrists as isothermal calorimeters, whereby the temperature of the surrounding (isothermal jacket) and the calorimetric vessel remains constant. The assumption here is that since the heat produced is transferred to the heat sink immediately, the temperature of the reaction vessel and the surrounding remains constant. However, for heat to be measured, it has to flow from a higher to a lower temperature gradient across the thermopile wall, thus, an ideally isothermal state can not be achieved (Hemminger and Sarge 1998). Therefore, in a strict sense, isothermal calorimeters are actually isoperibolic, whereby the surrounding has a constant temperature, with temperature of the measuring system possibly varying from it.

The construction principle of calorimeters could be a *single measuring system* or a *twin/differential measuring system*. The heat conduction types of calorimeters are generally constructed on the twin/differential principle, and are essential when high precision is required for slow process microcalorimetry (Kemp 1998). The two vessels of the twin setup (the reference and reaction vessels) are arranged as perfect twins with the detection units being in opposition, in order to give a differential signal. Thus, extraneous disturbances are cancelled giving long-term stability and precise results.

Isoperibolic calorimeters are among the most important types of calorimeters used in the investigations of living systems without interfering with their physiological demand. Such calorimeters are important in biological investigations because most biological reactions have a narrow optimum temperature range which can not be maintained by the other calorimetric types. The calorimeters used in this thesis work are categorized as *isoperibolic differential heat conduction microcalorimeters*; their working principle and interpretation of signals shall be considered in the next section.

1.4.2 Isoperibolic heat conduction microcalorimeters

Isoperibolic heat conduction microcalorimeters involve the transfer of heat produced in the reaction vessel to the surrounding heat sink (isothermal jacket), due to temperature difference across a thermopile wall (sensor of heat flow) placed between the vessel and the surrounding (Wadsö 2002).

The rate of heat evolution in the reaction vessel (rate of heat change) is $P = dQ/dt$, and is measured in units of watt (W). This rate of heat change is sometimes called "thermal power", but this remains disputed (Gnaiger 1993). Part of the heat evolved in the reaction vessel is exchanged with the surrounding, and a part of it is contained in the sample and in the reaction vessel that contains it. The sum of both is equal to the rate of heat change

$$P = dQ/dt = \Phi + C*dT/dt \qquad (1)$$

where Φ is the heat flow (rate of heat exchange) between the reaction vessel and the surrounding. The term $C*dT/dt$ represents the rate of heat accumulation in the reaction vessel. C is the heat capacity of the sample and the reaction vessel system (including part of the measuring sensors), and dT/dt is the rate of change of the temperature of the reaction vessel. The heat flow rate from the reaction vessel to the surrounding, Φ, is directly proportional to the temperature difference between the reaction vessel and the surrounding

$$\Phi = K(T_{sample} - T_{surrounding}) \qquad (2)$$

where K is the coefficient of thermal conductivity between the reaction vessel and the surrounding. By combining equations 1 and 2 above, the rate of heat change, P, in the sample is related to the temperature in a calorimetric system as given by equation (3)

$$P = K(T_{sample} - T_{surrounding}) + C*dT_{sample}/dt \qquad (3)$$

The ratio between the total heat capacity of the measuring system C, and the heat exchange coefficient between the sample and the surrounding is the time constant of the calorimeter, τ, and is very important to consider for reaction kinetics where it is necessary to observe the beginning and/or end of a reaction.

$$\tau = C/K \qquad (4)$$

The total quantity of heat, Q, evolved in the reaction vessel, during a given experimental period can be determined by integrating equation 3:

$$Q = \int P = \int_{start}^{end} \Phi dt + C\Delta T \qquad (5)$$

The heat conduction calorimeters measure the heat flow, Φ, between the reaction vessel and the surrounding, thus the second term in equation 3 becoming a dynamic correction factor. For slow reactions, like the biological systems investigated in the present case, and for close to perfect heat conduction calorimeters, the $C*dT_{sample}/dt$ is insignificant and equation 3 becomes

$$P = \Phi \qquad\qquad\qquad (6)$$

In (nearly) perfect heat conduction calorimeters the rate of temperature change in the sample and reaction vessel, ΔT in equation 5, is zero and integration of measured heat flow Φ gives the heat quantity, Q.

1.4.3 Calibration of calorimeters

Calorimeters can be calibrated by different means, such as by using the transition enthalpies of known reference materials, specific heat capacity, or direct electrical calibrations (Haines et al. 1998), or heat of chemical reactions (e.g. hydrolysis, and neutralization) of selected compounds (Briggner and Wadsö 1991, Kemp 1998, Beezer et al. 2001, O'Neill et al. 2003). Except for two of the Calvet calorimeters, where external resistors were employed, all calorimeters used in this study have built-in calibration heaters of known resistance. For this reason, and also since electrical calibration is a convenient method that can be done routinely as often as possible, calibration of all calorimeters was carried out electrically. For those calorimeters with no built-in calibration resistor, electrical calibration was done by passing precisely controlled electrical current by means of a constant current supplier (Electro Automatic EA, Viersen, Germany), through a calibration heater with a resistance of 124 Ω.

Electrical calibration can be done by passing a current, I (A) of known quantity through a resistor and recording the calorimetric output/thermopile potential, U (V) which is directly proportional to the temperature difference between the reaction vessel and the surrounding heat sink. The power input P is represented by

$$P = I^2 R \qquad\qquad\qquad (7)$$

where I is the current input (A) and R is the resistance (Ω) of the calibration heater. The sensitivity of the calorimeter S (V/W) is the ratio between the thermopile potential U (V) and the power input of electrical energy P (W). For nearly perfect heat conduction calorimeters the power input P is the same as the heat flow rate from the reaction vessel to the surrounding (Φ) as displayed in equation 6. Thus,

$$S = U/\Phi \qquad\qquad\qquad (8)$$

Calibration has to be carried out under conditions as close as possible to those of the sample experimental run. The flow calorimeters, especially, have to be calibrated (electrically) during pumping the sterile growth medium (in case of microbial investigations).

1.4.4 Advantages and disadvantages of calorimetry in biological investigations

Biological calorimetry is a general, non-specific method that measures the net enthalpy change that results from the complex metabolic reactions of living systems. The advantages and drawbacks of biological calorimetry both lie in the fact that measurement of heat flow is unspecific (Lamprecht 1983). The advantage of its non-specificity is that it monitors all heat producing reactions and, hence, can detect unexpected life processes that could be overlooked by other more specific methods (Lamprecht 1983, Wadsö 2002). The non-specific calorimetric signal from a complex biological reaction is unfortunately difficult to interpret at a molecular level in the absence of more specific analytical information. This difficulty can, however, be solved by equipping the calorimetric vessel (or the flow line in case of flow microcalorimeters) with specific analytical sensors such as electrodes and spectrophotometers, making the setup a very powerful analytical instrument for the interpretation of complex biological reactions (Johansen and Wadsö 1999).

Advantages of the calorimetric method compared to other techniques include its non-invasive nature - measuring heat production without interfering with the organism, and no need of clear solution – unlike spectrophotometric methods. In addition to this, the calorimetric method has a higher sensitivity compared to most standard methods, such as in the investigation of sublethal effects of toxicants on the metamorphosis of insects.

The position of the reaction vessel, deep in the calorimetric chamber, makes it difficult to mix the contents, and to supply essential materials such as oxygen in batch calorimeters, because stirring and/or pumping can introduce artefacts in the calorimetric signal (Lamprecht 1983). These problems can be minimized by the use of flow calorimeters, but the problem of exhaustion of oxygen in the flow line still persists, especially at higher cell densities and hence interpretation of results has to be done with caution and in combination with signals from oxygen sensors incorporated in the flow line.

1.5 Objectives and structure of the thesis

Chapter 1: As aforementioned at the beginning of this chapter, it is important to give general introductions about Varroa mites and varroosis, wax moths, propolis, and calorimetry. Thus, in this chapter the biology of *Varroa destructor* mites, the extent of problem of varroosis, possible means of coexistence of some resistant honeybee species/subspecies with the mite, by controlling the population size below a certain threshold are demonstrated. A brief insight into the biology and infestivity of *Galleria mellonella* is made in order to show the extent of the problem. In addition, this chapter also gives clues as to the what-about of propolis, its chemical make-up, botanical origin, and collection and use by the honeybees. The last part of this chapter deals with the important technique, calorimetry, used in most investigations of this thesis research. The calorimetric topic demonstrates the working principles, sensitivities compared to other methods, reliability, and calibration and standardization of calorimeters.

Chapter 2: It is still highly controversial among bee researchers and beekeepers, as to whether the energy and nutritional demand of *Varroa destructor* mites and, thus, the amount of hemolymph they suck from their host (mainly brood), is by itself responsible for the weakening and consequential death of honeybees. Most of the hitherto evaluations of the energetic and nutritional demand of Varroa mites are more of speculations rather than experimental proofs. Several researchers demonstrated that *Varroa destructor* mites transmit viral infections. The transmitted viruses were considered to play primary roles in killing bees, whilst the mites playing a secondary one, or just simple vectors with no much impact on the bees. This confusion and debate about the role of *Varroa destructor* arises mainly due to undermining the amount of hemolymph they suck from brood. Thus, this chapter is devoted to the experimental and computational proofs of the amount of hemolymph mites suck, and their energy demand from a capped brood. Calorimetric methods are used in the investigations of the energy demand of Varroa mites and brood during the capped developmental stage. The amount of hemolymph the mites suck during brood development is evaluated by starving the mites, incubating them in the absence of their host, and measuring the weight loss. The logic behind measuring weight loss of starving mites is that had it not been for the absence of their host, they would have sucked an amount of hemolymph equal to the lost weight and maintained their weight constant. The length of time spent by each individual (mother and offspring) mite in a capped brood will be taken into consideration for the computation of mites' energy and nutritional demand.

Chapter 3: Varroa destructor has become a global problem of the beekeeping industry based on *A. mellifera*. In order to stop the weakening and consequential death of colonies, beekeepers are treating them with acaricides. The use of acaricides, however, is associated with residue problems in bee products and mites resistant to currently used acaricides have already emerged. These problems provide incentives to search for new and potential acaricides which are free from the mentioned problems. Natural products are becoming the subject of such investigations. This chapter deals with the *in vitro* investigations of the acaricidal action of propolis. The Varroa weakening and varroacidal actions of various concentrations of propolis extracted in 70% or 40% ethanol are evaluated with different lengths of contact times. The screening for optimum concentration and contact time may help in the future development of treatment method *in vivo*. Evaluation of the effects of various concentrations and treatment times are carried out by counting the number of dead/inactivated mites at different time intervals, and through calorimetric monitoring of the heat production rates of mites before and after treatment with propolis.

Chapter 4: The use of high temperature treatment (e.g. 44 °C) to differentially kill the infesting mites in a capped brood was found out to be effective with little impact on the latter. One of the drawbacks of high temperature treatment is the length of exposure time, leading to brood/bee death. It is, therefore, desirable to develop a method of shortening the treatment period. One of such methods would be the exploitation of the synergistic effect of lower to moderately concentrated acaricides and high temperature treatments. This chapter concentrates on the calorimetric and respirometric investigations of the effect of temperature on the antivarroa actions of propolis. Investigations are carried out on mites collected from drone brood, worker brood, and adult workers separately, since they may have different responses. It is usually recommended that calorimetric results have to be supported by other data, to concretely explain the changes that take place in the experimental organism. The calorimetric results at different temperature setups are compared with those of manometric experimental results.

Chapter 5: Propolis samples from different geographic origins were shown to be highly variable in chemical composition. It is not clear whether this variation in chemical composition affects the antivarroa action of propolis or not. Comparisons of the antivarroa action of propolis of different geographic origins are made in this chapter. Apart from samples of different geographic origins, the antivarroa actions of propolis samples from the same location, and even from the same apiary but different beehives are compared. The species/subspecies of bees that collected propolis are considered whilst comparing the antivarroa actions of different samples. The differences in the antivarroa actions of various extracts of the same sample are also made.

Chapter 6: Apart from *Varroa destructor,* the beehive harbours several parasites and pests that cause enormous loss, and have to be controlled. One of such pests is the greater wax moth *Galleria mellonella.* Beekeepers treat their colonies with insecticides to save them from death, but insecticides used against wax moths cause residues in bee products, and may irritate bees and beekeepers. One of the best solutions to the problems associated with synthetic insecticides is the use of natural products. This chapter deals with the *in vitro* investigation of the effects of propolis on the metabolism and development of the different developmental stages of *Galleria mellonella.* In a first group of experiments the effect of propolis on the heat production rates of the different larval stages are investigated to evaluate the change of sensitivity to propolis treatment with changing larval instars. Since the treatment with lethal doses of propolis does not make sense for calorimetric experiments, most of the calorimetric investigations are carried out with sub-lethal doses. Apart from the reduction of heat production rate of an organism, by weakening it, some plant secondary metabolites were observed to play the roles of insect growth regulators, either by facilitating or retarding larval and/or pupal development. The second group of experiments, thus, concentrates on the effect of sublethal doses of propolis on pupal metamorphosis. This is achieved by treating the last larval instar of Galleria *mellonella* with propolis, and calorimetrically monitoring the events of metamorphosis up to adult emergence. Different events and parameters of metamorphosis, like the strength of endothermic trough and exothermic peaks during ecdysis, length of the metamorphotic period, the basic metamorphotic heat production rate, rate of pupal reserve food utilization, etc., are evaluated.

Chapter 7: Propolis is claimed to be a multifaceted drug against various types of bacterial and fungal infections. Several investigations were carried out in different laboratories to proof its potential, but most of them have one or more limitations. Insight into possible solutions of the existing limitations of propolis antimicrobial research, new queries and problems are dealt with in this chapter. The majority of hitherto investigations were based on only one type of extract, mainly the ethanol-extracted propolis, and experiments with other types of extracts are rare. The use of aqueous solutions of propolis, however, could be desirable under certain circumstances, especially for human medicinal use. Therefore, the antibacterial and antifungal activities of three types of extracts of propolis, namely the ethanol-extracted propolis (EEP), water-extracted propolis (WEP), and Propolis volatiles (PV) are investigated and compared in this chapter. In addition, comparison of the antimicrobial activities within, and between different extracts is made in relation to some physico-chemical parameters, and the yield of propolis extraction.

Almost all of the hitherto antimicrobial investigations of propolis were carried out using the Petridish bioassay method which is actually a highly constrained method for the investigation of the mechanisms of action of antimicrobial agents, especially the hydrophobic ones. In this chapter, the mechanisms of action of propolis are investigated by means of flow microcalorimetry coupled with polarographic oxygen sensors. Results from the flow microcalorimetric investigations are compared with those obtained from the Petridish bioassay methods.

Up to now most propolis samples used by researchers were usually obtained from a certain geographic region. Comparisons of samples from regions of completely different climatic zones are rare. Hence, this chapter deals with various extracts of propolis samples obtained from geographic regions of different climatic zones and vegetation compositions, in order to observe the effect of geographic origin on the activity of one or more of the extracts.

Chapter 8: A general discussion of the extent of problems of honeybee parasites and pests, currently existing solutions to these problems and their drawbacks are given in this chapter in English and German languages. In addition, the findings of the different chapters of this thesis work, and their potential *in vivo* applications in the control of such parasites and pests are dealt with.

Chapter 9: This chapter gives short summaries of the thesis in German and English languages.

Chapter 10: Here the emphasis is on planned future work, and what has to be done before propolis is to be used in the control of honeybee parasites and pests.

Chapter 11: Contains the list of reference materials used during the research phase and while writing the thesis.

Chapter 12: Is a compilation of appendix, personal data, a list of my own publications, and declaration.

2.0 Metabolic Physiology and the Energy and Nutritional Demands of the Parasitic Life of the Mite *Varroa destructor*

2.1 Abstract

The energy and nutritional demand of the ectoparasitic mite *Varroa destructor* (Anderson and Trueman) was investigated by calorimetry, respirometry and their resource utilization rate. Mites from worker brood, drone brood, and adult workers of the western honeybee *Apis mellifera carnica* were monitored in the absence of the host. Energy metabolism of the mites, elucidated by the heat production rate, was insignificant as a factor to cause honeybee colony death. The metabolic rates of mites ranged from 1.1% to 2.4% of that of the bee pupae, depending on the infestation level. The nutritional demand of the mites was, however, very high, owing to their inefficient metabolic machinery, utilizing up to 28% of reserve food of the pupa during the capped brood developmental stage. This contributes to the malformation and weakening of the bees and eventually death of the colony.

2.2 Introduction

The formidable ectoparasitic mite *Varroa destructor* (Anderson and Trueman) feeds on the hemolymph of the brood and adults of the western honeybee *Apis mellifera* L. and causes damage to the latter. The extent of harm is proportional to the degree of infestation. Varroa mites may involve different mechanisms in the weakening and obliteration of the honeybee colony. The role of *Varroa destructor* as a vector of bacterial and fungal disease as well as an inducer of latent viral infection in varroosis of the western honeybee *A. mellifera* L. has been demonstrated (Hargasm 1973, Ball 1983, Ball and Allen 1986, Wiegers 1986, Trubin et al. 1987, Gliński and Jarosz 1988, 1990a, 1990b, 1992, Ball 1994 and 1996, Liu 1996). Ball (1983 and 1996) mentioned that Varroa mites cause open wounds on the surface of the bees or brood, through which viral invaders get access to the hemolymph. These viral infections are commonly suspected to be the primary cause of bee mortality in colonies of *A. mellifera* severely infested with *V. destructor*. Destruction of tissue and impairment of the immune system of the host by the parasite can also induce the development of latent viral infections by releasing infectious agents from damaged tissues and stimulating replication of the viral infectious agent in newly infected honeybee cells (Wiegers 1986, Gliński and Jarosz 1984, 1988). Apart from its role as a vector of bacterial, fungal and viral diseases, the damage caused by infestations even with a few number of mites on the ontogenesis and weight at hatching of the honeybee *A. mellifera* was demonstrated by Schneider and Drescher (1987). The researchers showed that the loss of weight of a bee infested at the brood stage, compared to non-infested bees, is directly proportional to the number of infesting mites. Infestation of the bee brood with a large number of mites results in the formation of adults with malformed wings, underdeveloped and short abdomen (Hargasm 1973, De Jong et al. 1982, Marcangeli et al. 1992), and underdeveloped hypopharyngeal glands (Schneider and Drescher 1987). It has also been demonstrated that, apart from the lower weight at hatching, honeybees parasitized with Varroa mites in their pupal stages have lower life spans than the unparasitized bees (De Jong et al. 1982, De Jong and De Jong 1983, Schneider and Drescher 1987). In general, bees that emerge from brood infested with higher number of mites are deformed, and incapable of normal life activities (Fig. 2.1).

Fig 2.1 Pre-capped worker brood infested with *Varroa destructor* (a), and a deformed adult worker that has been infested with Varroa mites during pupal development (right), and a non-infested healthy one (left) (b).

Investigation of the chemical composition of the hemolymph of pupae infested with Varroa mites displayed lower protein concentration than in the unparasitized ones (De Jong et al. 1982, Weinberg and Madel 1985, Kovac and Crailsheim 1988, Bailey and Ball 1991). The reduced protein concentration in the hemolymph may explain why some organs of the honeybees infested in the brood stage are malformed or underdeveloped, since proteins are generally important in organ formation during ontogenesis. Bees that were not infested with mites at the brood stage but artificially infested immediately after hatching showed less developed hypopharyngeal glands than the non-infested bees (Schneider and Drescher 1987), demonstrating the energy and nutritional demand the parasite imposes on the immature adult host. In turn this leads to underdevelopment of the bee. Infestation of drone and worker pupae with equal numbers of mites resulted in similar weight losses at hatching, demonstrating that the nutritional value of the hemolymph from worker and drone pupae is the same for the mites (Schneider and Drescher 1987, De Jong et al. 1982).

Apart from the indirect evidence and notion that the mites can be the secondary cause of honeybee colony death, the primary cause being the viral infections transmitted by the mites, no data exists on the energy metabolism of the mites. In this work the energy and nutritional demands of *V. destructor* from various developmental stages and sexes of the honeybee *A. mellifera carnica* shall be demonstrated calorimetrically and gravimetrically at different experimental temperatures. The implications of these demands on the vigour and activities of honeybees will then be evaluated. Furthermore, the rate of utilization of reserve food by the mites during starvation, and also the length of time the mites could survive starvation in the absence of their host shall be examined.

2.3. Materials and Methods

2.3.1 Evaluation of brood infestation level

In order to compare the degree of infestation of worker and drone broods of the honeybee *A. m. carnica* by the mite *V. destructor,* and to subsequently evaluate the energy and nutritional demand of the mites, infested honeycombs were obtained from the research beehives of the Institute of Zoology, Free University of Berlin, Germany in summer 2001. Fifteen honeybee combs containing both types of broods, the worker brood located centrally and drone brood peripherally, were used in the brown and dark skin pupal stages. The combs with brood were collected randomly from the central part of the hives of five colonies, three from each. The experimental colonies were treated with formic acid only at the beginning of autumn of the preceding year. From each comb 200 brood cells (100 from each side) of each sex were selected randomly, opened carefully under a binocular microscope, and all the developmental stages of the female mite were counted

2.3.2 Calorimetric determination of the metabolic rate of capped worker brood

Metabolic rates of the different developmental stages of the capped worker brood of the honeybee *A. m. carnica* were investigated by using isoperibolic twin heat conduction calorimeters. THERMANALYSE (Messgeräte Vertrieb, München, Germany) and Calvet calorimeters (MS 70, Setaram, Lyon, France) of vessel volumes of 15 or 100 cm^3, and sensitivities between 40.65 and 64.34 μV mW^{-1} were used. Capped worker brood was taken out of the hive immediately after capping ± 12 h, and incubated further at 35 ± 0.5 °C and 65 ± 5% r.h. A capped brood cell at a certain developmental stage was carefully incised from the rest of the brood, put into a calorimetric chamber, and the heat production rate was recoded for 3 to 5 h. The brood cell was then opened, and the pupa was removed and weighed using a sensitive analytical balance (Type 414/13, Sauter, Ebingen, Germany) of 0.1 mg detection limit. Experiments with each developmental stage were done 12 to 15 times and the mean ± S.D. values were used in the presentation of results.

To avoid the use of Varroa infested and/or diseased brood, the bees were optically examined through the incised transparent brood cell. Brood cells that were infested with Varroa mites and pigmented differently, possibly due to infection, were excluded. Since observation through an old and dark wax is difficult, only brood on new and clear combs were used in the present experiments. In addition to the preliminary observations, brood cells were opened and inspected at the end of each experiment and those with the above mentioned problems were discarded.

2.3.3 Determination of weight change of brood during ontogenesis

In order to determine the rate of weight change and hence reserve food utilization by the capped brood, 15 brood cells that were capped in the last 12 h were sliced carefully from the rest of the brood and weighed. The individual brood cells were then incubated at 35 ± 0.5 °C and 65 ± 5% r.h. in a Petridish. Each of them were weighed every 24 h until hatching. At the end, the weight of the individual empty comb cells from which the adults hatched was determined separately, and subtracted from the previous weightings. The weighing times were maintained as short as possible, < 1 min, in order to avoid the effect of lower room temperature and humidity on pupal development.

2.3.4 Survival of starvation and resource utilization by *Varroa destructor*

In order to indirectly determine the amount of hemolymph the mites suck from their host, and to elucidate the length of time the adult female mites could survive in the absence of their host, mite starvation experiments were carried out. For these tests 130 to 190 mites were collected from brown and dark skin drone pupae (brood older than 20 days), weighed using the analytical balance (Sauter, Ebingen, Germany) and incubated at 35 °C and 60 ± 5% r.h. for 1, 6, 12, 18, 24, 30, or 36 h in a Petridish with or without pupae, approximately 4 to 6 mites per pupa in the former case. At the end of each experiment, the mites were weighed again, and immediately frozen for further analysis. For the determination of the change of body fat and dry matter composition with starvation time the frozen mites were thawed and dried at 60 °C for 96 h until weight constancy, and the mean of the final three weightings, performed every 12 h, was used in the calculation of percentage dry matter. The dried samples were then homogenised using mortar and pestle. Analysis of the fat composition was done using the Folch method (Folch et al. 1957). Each of these experiments was conducted three times and the mean ± S.D. values were used in the presentation of results.

2.3.5 Calorimetric experiments with *Varroa destructor*

Calorimetric experiments were performed to determine the metabolic rate of *V. destructor* and to evaluate its energy demand from the host. Isoperibolic twin heat conduction calorimeters of types: Biocalorimeter - B.C.P (Messgeräte Vertrieb, München, Germany) with a sensitivity of 44.73 μV mW^{-1} and a vessel volume of 12 cm^3, and a THERMANALYSE calorimeter (Messgeräte Vertrieb, München, Germany) with a sensitivity of 40.65 μV mW^{-1} and a vessel volume of 15 cm^3 were used.

To evaluate differences in the energy metabolism of mites from different sexes and developmental stages of the same sex, mites were collected from adult workers as well as worker and drone brood cells. Mite collection was done from the brood stage by gently opening individual healthy (non-infected) brood cells. During the collection process mites were kept in a Petridish on the corresponding bee pupae in order to avoid starvation. Newly moulted adult mites, identified by their pale colour, and young mites (nymphs), with relatively smaller size and feeble locomotion, were excluded from the experiment. Mites that seemed weak and abnormal were discarded. The collection of mites from adult workers was done by very carefully dislodging them from the surface of the bees with the help of a blunt needle. During the collection of mites from adult workers the former were kept on worker pupae in order to avoid starvation.

25 to 30 mites were weighed and put into the calorimeter and the heat production rate was recorded for 2 to 3 hours. This time interval was chosen for technical reason with the calorimetric experiments, due mainly to the thermal equilibration time needed after opening the calorimeter and placing the mites inside. The thermal equilibration time of the calorimeter was not always the same and varied based on several factors. The little difference in the experimental time interval for the different experiments does not affect the result since the latter is extrapolated to a rate per hour. At the end of each experiment the mites were weighed again for weight change and evaluation of the rate of utilization of reserve food under starving conditions, and consequently to elucidate the amount of hemolymph the mites could utilize from their host

to maintain their weight and physiological condition. The weight change is presented as percentage wet weight change per mite per hour.

All calorimetric and gravimetric experiments with mites from the different groups were conducted at 35 °C. In addition, mites from drone brood were investigated at 25, 30, 35, 40, 45, and 50 °C in order to determine the optimum temperature of mite metabolism and to elucidate the effect of temperature on their metabolic rate and rate of resource utilization. The experimental time with mites at 45 °C and 50 °C lasted 1.5 to 2 h.

2.3.6 Respirometric experiments with *Varroa destructor*

The respiration rate of mites from drone brood was determined at 25, 30, 35, 40, 45, and 50 °C using manometric methods. The tests were run in Warburg vessels of about 12 ml volume. The CO_2 evolved was absorbed by 400 μl of a 4% KOH solution in the side arm of the vessel. To avoid access of the mites into the side arm the opening was blocked with a very thin layer (1 mm thick and 0.8 mm pore size) of porous spongy material that allowed air, but prevented mite entrance. The measurements were started after a temperature equilibration time of 30 min and recording was done in intervals of 30 min for 3 to 5 h (except for 1.5 to 2 h at 45 and 50 °C). The oxygen consumption rates were calculated from the pressure drop in the Warburg vessel with time. Each measurement was carried out five times (but nine times at 45 °C) using 50 to 60 mites per experiment, and the mean ± S.D. values were used in the presentation of results. The latter were compared with those of the calorimetric ones, also at different temperature settings.

Finally, the metabolic rate of the mites obtained from different developmental stages of honeybees was compared with that of the corresponding developmental stages of the host from which the mites were collected. Furthermore, the contribution of the energetic demand of mites on the deterioration and death of infested colonies was evaluated.

2.3.7 Statistical analysis

Results are presented as mean ± S.D. values. Statistical tests were performed using the student's t-test, 1-way ANOVA, and the Tukey's HSD post hoc test. $\alpha = 0.05$ was considered as the critical value.

2.4 Results

2.4.1 Infestation level of brood

Comparison of the infestation level of broods of the same sex located on the two sides of a comb and also from combs of different colonies showed no significant differences. Due to this fact the results were pooled and presented as infestation level of worker brood or drone brood. As can be seen from Fig. 2.2 there is a significant difference in the percentage of infested worker and drone brood cells, the mean values being 5.6 ± 0.8% and 28.4 ± 7.5%, respectively. These results illustrate that drone broods are fivefold more attractive to mites than worker brood (drone brood preference factor). A significant difference between drone and worker brood was also observed in the mean number of female mites counted per 100 brood cells. The 100 drone brood cells opened randomly on each side of the comb and from combs of each colony rendered a mean value of 113.4 ± 13.5 female mites. A corresponding number of worker brood, however, rendered only 7.7 ± 1.4 mites, indicating that the infestation level of the drone brood by female mites is significantly greater than that of the worker brood by 14.8 fold. If we consider individual infested brood cells, a drone brood harbours 3 to 6 female mites with a mean value of 4.1 ± 1.0, whereas a worker brood harbours 1 to 4 female mites with a mean value of 1.4 ± 0.8, indicating that an individual drone brood shelters female mites more than a worker brood by a factor of 2.9.

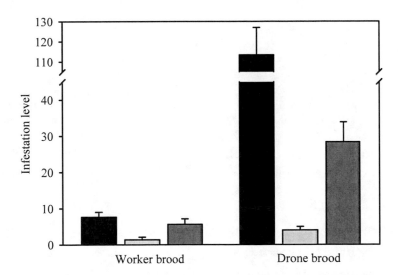

Fig. 2.2 Infestation level of *Apis mellifera carnica* worker and drone brood with *Varroa destructor*. ■ Average number of female mites per 100 brood cells, ▢ Average number of female mites per infested cell, ▨ Percentage of infested brood cells (%).

2.4.2 Metabolic rates and utilization of reserve food under starvation

The starvation experiments with mites from drone brood in the absence of the host showed that no death of mites was observed at least within the first 6 hours. Death ensued with prolongation of the starvation time with about 22% of the mites dying within 12 h, and 50% within 18 h of starvation time. Only 5% of the mites were alive after 36 h starvation time. There was neither death nor loss of weight of mites incubated on drone pupae (control experiment) at least for the 36 h experimental time. The starving mites lost 48.5% of their wet weight within the first six hours of starvation time, which is a loss of 8.1% h^{-1}. The wet weight change of Varroa mites with starvation time during the calorimetric experiments for 2 to 3 hours also displayed a loss of 8.1% h^{-1}. This trend of weight loss did not continue further in the next starvation hours, rather it reduced during the course of incubation time (Fig. 2.3). After 36 h starvation time, when 95% of the mites had already died, the wet weight was reduced by 76%.

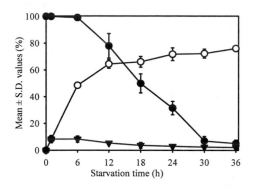 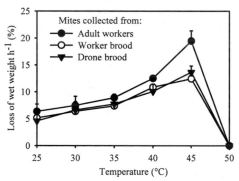

Fig. 2.3 Survival of *Varroa destructor* mites and the rate of reserve food utilization under starvation in the absence of their host, incubating the mites at 35 °C and 60 ± 5% r.h. in a Petri dish. ● Mean percentage of survivor mites (%), ○ Mean percentage loss of wet weight (%), ▼ Mean percentage loss of wet weight per starvation hour (%).

Fig. 2.4 Percentage loss of wet weight (mean ± S.D.) per hour of the mite *Varroa destructor* under starvation and different experimental temperatures, data extrapolated from starvation of the mites for 2 to 3 hours at the corresponding temperature. n = 5, 25 to 30 mites per experiment.

Extrapolation of the wet weight loss of a mite per hour from the weight loss during the 2 to 3 h starvation time in the calorimetric experiments displayed that the percentage weight loss grew with increasing temperature and attained maximum values at 45 °C, amounting to 13.7 ± 1.1%, 12.5 ± 0.6%, and 19.5 ± 1.9% mite^{-1} h^{-1} for mites from drone brood, worker brood and adult workers, respectively. Activity of mites at this temperature lasted a maximum of 3 hours, and thereafter mites were either dead or highly weakened. The weight change at 50 °C dropped to zero since the mites died immediately. If we consider the weight loss at 35 °C, the temperature prevailing in the beehive environment, mites from drone brood, worker brood and adult workers lost $8.1 \pm 0.6\%$, $7.8 \pm 0.9\%$, $8.9 \pm 0.7\%$, respectively, of their wet weight per hour (Fig. 2.4). A one-way ANOVA showed that there is a significant difference in the loss of weight of mites from the different sources ($p = 0.035$). A pairwise comparison using the Tukey's HSD post hoc test displayed a significant difference of weight loss by the mites from drone brood and adult workers ($p = 0.013$), and worker brood and adult workers ($p = 0.02$), but not by the mites from drone brood and worker brood ($p = 0.25$). The mites incubated at the different experimental temperatures and on the corresponding hosts (controls) did not show any weight change with the incubation time. The weight of mites collected from the three groups of bees in their normal physiological conditions and before starvation did not show any significant difference, even though those from adult workers were slightly lighter, the values being 407 ± 38, 395 ± 43, 361 ± 63 µg for mites from drone brood, worker brood and adult workers, respectively. This could be due to the fact that the brood mites were collected from the late pupal stages (brown and dark skin stages) when there is no much oogenesis, which otherwise would have made the brood mites heavier by up to 30% (Steiner et al. 1995).

The chemical analysis results of starving mites showed that the mean percentage of dry mass remained constant and similar to the control experiments at 68% regardless of the starvation time. In addition, the relative proportion of the lipid to non-lipid components of the dry mass also remained unaffected with the lipid components varying between 11.5% and 14.8% (mean $13.2 \pm 1.6\%$), and the non-lipid components lying between 84.8% and 88.0% (mean $85.3 \pm 5.9\%$) independent of the starvation time. Observation of the inner wall of the transparent calorimetric vessels after each experiment displayed several white spots on the otherwise clean Plexiglas vessel. These white spots are faecal matter of the mites which are considered by Sammataro (et al. 2000) to be mainly composed of guanine.

Calorimetric experimental results at 35°C indicated that mites collected from worker brood and drone brood had comparable (specific) heat production rates, expressed either as µW mg^{-1} or µW $mite^{-1}$. Though 1-way ANOVA showed no significant difference ($p = 0.45$) in the heat production rate per individual mite (µW $mite^{-1}$) between the mites collected from adult workers, drone brood, and worker brood, the same test showed a significant difference ($p = 0.024$) in the specific heat production rates (µW mg^{-1}) of these mites. The Tukey's HSD post hoc test displayed that significant difference exists in the specific heat production rates between mites from drone brood and adult workers ($p = 0.017$), and those from worker brood and adult workers ($p = 0.023$) but not those between worker brood and drone brood ($p = 0.35$) (Fig. 2.5).

Calorimetric and respirometric results of the effect of temperature on the metabolic rate of Varroa mites from drone brood showed identical patterns. Heat production and oxygen

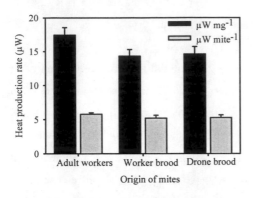
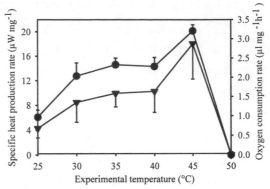

Fig. 2.5 (Specific) heat production rate (µW mg^{-1} or µW mite^{-1}) of *Varroa destructor* mites collected from different sexes and developmental stages of the honeybee *A. mellifera carnica* at 35 °C. 25 to 30 mites per experiments, n = 5.

Fig. 2.6 Effect of temperature on the specific heat production rates and oxygen consumption rates of *Varroa destructor* mites from drone brood. 25 to 30 and 50 to 60 mites per experiment for the calorimetric and respirometric experiments, respectively. n = 5 (but n = 9 for the respirometric measurements at 45 °C), mean ± S.D. ● Heat production rate, ▼ Oxygen consumption rate.

consumption rates of mites were small at lower temperatures and rose with increasing temperatures reaching optimum values between 30 °C and 40 °C. In this temperature range both curves attained nearly a plateau phase demonstrating little change of metabolic rate with changing temperature. If we compare the Q_{10} values between 25 °C and 35°C on one hand, where there is a drastic change of heat production rate, and between 30 and 40°C on the other hand, the Q_{10} value of 2.4 for the former is significantly higher than for the latter, which is 1.1 (two-tailed t-test, p = 0.009). Both heat production and oxygen consumption rates increased and achieved maximum values at 45 °C. These elevated metabolic and oxygen consumption rates at 45 °C were, however, maintained only for a maximum of 3 h, after which the mites started dying, and the rates dropped drastically (Fig. 2.6).

2.4.3 Computation of the brood hemolymph robbed by a foundress mite and its offsprings

In order to calculate the energy and nutritional demand of the mites from the brood and to depict the effect of parasitism, the following facts have to be taken into consideration in combination with data obtained in this work: a foundress mite invades a worker brood on average ca. 20 h and a drone brood 45 h before cell capping, and starts feeding on the brood's reserve food, and later on the brood's hemolymph (Boot et al. 1991). The post-capping period of the brood lasts 12 and 15 days for worker and drone brood, respectively (Moritz 1985, Le Conte and Cornuet 1989). The developmental time of the mite from egg to maturity is 6.9 and 6.2 days for males and females, respectively (Rehm and Ritter 1989). The first egg is laid 60 h after cell capping and develops into a male mite and the rest follows in intervals of 30 h and develops into female mites (Infantidis 1983). Based on these facts the feeding time of the mother mite and its offsprings is tabulated in Table 2.1. The energy and nutritional demand the mites impose on the brood (Table 2.2) is calculated from (i) Table 2.1, (ii) the daily energy demand of the mites (extrapolated from Fig. 2.5) and (iii) the amount of hemolymph the mites suck from the brood (extrapolated from the loss of weight in the absence of the host). If the foundress mites are two instead of one, the energy and nutritional demands of the mites on the brood change, with both values increasing as depicted in Table 2.3.

The metabolic rates of the capped brood recorded calorimetrically, and that of the infesting mites calculated at each instar based on the number of mites and the mean heat production rate of 4.8 μW mite^{-1} were compared (Fig. 2.7). Though the energy consumption rate of Varroa mites increases with the incubation period of the brood, as seen in Fig. 2.7, it is a non-significant factor to cause damage to the brood since it is about 1.2% to 2.8% of that of the heat production rate of the latter.

The nutritional demand of Varroa infestation was also compared by considering resource utilization of the capped brood and its infesting mites in the capped developmental days. Weight of a worker larva achieved its maximum value at the L6 (the capped 5th instar larva on the 6th day of larval development) with 182 $\pm$ 17 mg. The weight achieved at this stage decreased continuously until adult emergence; with a total loss of 77 $\pm$ 9 mg for a non-infested brood, which is about 42% of the maximum weight (Fig. 2.8).

The rate of reserve food utilization by the developing brood decreased drastically in the first two days after capping, i.e. during growth from L6 to L7 and from L7 to PP1 (prepupal phase). After this period the rate remained more or less constant. The rate of robbing of brood's reserve food by the mites remained at a lower level during the first five capped developmental days (L6 to P1) and increased from the second to the sixth pupal developmental days (P2 to P6), where most of the progenies become mature. After the fourth pupal developmental day the mites consume up to 65% of the reserve food that the brood could consume Fig. 2.9

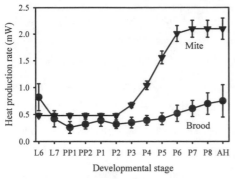

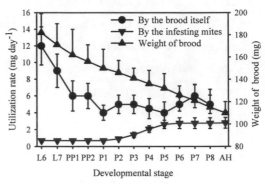

Fig. 2.7 Comparison of the heat production rate of a capped worker brood of the honeybee *Apis mellifera carnica* with an infesting *Varroa destructor* mother mite and its three offsprings during pupal ontogenesis. L = larva, PP = prepupa, P = pupa, AH = adult before hatching. n = 12 to 15 experiments for brood and n = 5 experiments for mites with 20 mites per experiment. The heat production rate of the mites displayed is a multiple of 10^2 of the actual value.

Fig. 2.8 Utilization of the reserve food of the developing capped worker brood of the honeybee *Apis mellifera carnica* by the brood itself and the infesting *Varroa destructor* mother mite with its three offsprings. L = larva, PP = prepupa, P = pupa, AH = adult before hatching. n = 12 to 15 for brood and n = 5 for mites with 20 mites per experiment.

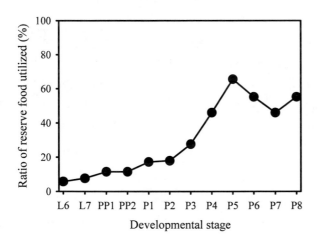

Fig. 2.9 Ratio of reserve food utilization by a mother *Varroa destructor* mite and its three offsprings to that utilized by the worker brood of *Apis mellifera carnica* during development in a capped cell.

Table 2.1 Feeding time of mites in a capped brood

The number of days a mother mite and its offsprings feed on worker and drone broods after invasion of the brood before cell capping. (* - The bee emerges as an adult before the mite completes development to the adult stage). The feeding of the first egg (male) is neglected.

	Egg number	Time (days) after cell capping at which egg is laid	Total time on brood (days)	Feeding time (days) of the mite neglecting the feeding period of the protonymphs i.e. the first 3 days
In a worker brood	Mother mite		12.8	12.8
	1	2.5	-	-
	2	3.8	8.2	5.3
	3	5.0	7.0	4.0
	4	6.3	5.7	2.8*
In a drone brood	Mother mite		17.0	17.0
	1	2.5	-	-
	2	3.8	11.2	8.3
	3	5.0	10.0	7.0
	4	6.3	8.7	5.8
	5	7.5	7.5	4.5
	6	10.0	5.0	2.0*

Table 2.2 Energy and nutritional demand of mites in a capped brood (1 foundress mite)

The total energy needed by a mother mite and its offsprings during the whole developmental time of a worker and a drone brood, obtained by multiplying the total feeding time with the energy demand of a mite per day: 471 mJ (from drone brood) and 491 mJ (from worker brood). The total weight of hemolymph robbed by the mites is a product of the total feeding time of the mites and their weight loss under starvation (extrapolated to hypothetical weight loss per day), which is 0.71 mg (from worker brood) and 0.72 mg (from drone brood). The energy and hemolymph demand of the first offspring (male) is neglected.

	The nth offspring	Total feeding time of the mites (days)	Total energy consumption of a mite (J)	Total weight of hemolymph and brood food consumed (mg)
In worker brood	Mother mite	12.8	6.3	9.1
	1	-	-	-
	2	5.3	2.6	3.7
	3	4.0	2.0	2.8
	4	2.8	1.4	2.0
			Total = 12.3	Total = 17.6
In drone brood	Mother mite	17.0	8.0	12.2
	1	-	-	-
	2	8.3	3.9	5.9
	3	7.0	3.3	5.0
	4	5.8	2.7	4.1
	5	4.5	2.0	3.2
	6	2.0	0.9	1.4
			Total = 20.8 J	Total = 32.8 mg

Table 2.3 Energy and nutritional demand of mites in a capped brood (2 foundress mites)
The amount of hemolymph sucked and the energy needed by two mother mites and their
offsprings during the developmental time of a worker and a drone brood considering that both
mother mites are fertile. Since the maximum infestation rate in the experimental colony was 4
and 6 mites per worker and drone brood, respectively, the evaluation is limited to the mother
mites, and the first female offsprings in the case of worker brood, and the mother mites and the
two successive female offsprings of each mother mite in the case of drone brood. FO1-1: first
female offspring of mother mite 1, FO1-2: first female offspring of mother mite 2, FO2-1:
second female offspring of mother mite1, FO2-2: second female offspring of mother mite 2.
Computation was done as in Table 2.2.

	The nth offspring	Total feeding time (days)	Total energy demand of the mite from the brood (J)	Total weight of hemolymph and brood food consumed (mg)
In Worker brood	Mother mite1	12.8	6.3	9.1
	Mother mite2	12.8	6.3	9.1
	FO 1-1	5.3	2.6	3.7
	FO 1-2	5.3	2.6	3.7
			Total = 17.8 J	Total = 25.6 mg
In drone brood	Mother mite 1	17.0	8.0	12.2
	Mother mite 2	17.0	8.0	12.2
	FO1-1	8.3	3.9	5.9
	FO1-2	8.3	3.9	5.9
	FO2-1	7.0	3.3	5.0
	FO 2-2	7.0	3.3	5.0
			Total = 30.4 J	Total = 46.2 mg

2.5 Discussion

Drone brood is highly infested compared to worker brood, and this phenomenon was also
confirmed by other researchers (Issa and Gonçalves 1984, Schulz 1984, Rosenkranz 1985, Fuchs
1990). Dividing the percentage of infested drone brood by the percentage of infested worker
brood gives a preference factor of 5 for drone brood to worker brood. Several factors could be
involved in encouraging the invasion of drone brood rather than worker brood cells. These
factors include higher concentrations of mite attractive fatty acid esters in the cuticle of drone
brood than in worker brood (Le Conte et al. 1989), higher concentration of aliphatic alcohols and
aldehydes in drone cocoons (Donzé et al. 1998), larger volume of the drone brood (Sammataro et
al. 2000), and longer Varroa attractive period prior to cell capping by drone brood than worker
brood (Fuchs and Müller 1988, Infantidis 1988, Boot et al. 1991). The number of female mites in
infested drone brood is higher than in worker brood by a factor of 2.9. This means that a
foundress Varroa mite reproduces more in drone brood than in worker brood, mainly because the
post capping period in drone brood is longer than that in worker brood by about three days
(Moritz 1985, Le Conte and Cornuet 1989). With the mean capped developmental time of 12

days for workers and 15 days for drones (Table 2.1), only the first two female mites mature to adults before the worker hatches, whilst four of the mites could mature before the drone hatches as an adult. Thus, female mites developing in the last eggs mature to adults if the capped brood has a longer developmental time than the average considered. The product of the preference factor of a drone brood to a worker brood and the ratio of infestation level of an individual drone brood to a worker brood gives the infestation ratio of drone brood to worker brood at the colony level, which is 14.8.

Starvation experiments carried out on mites to show how fast they consume their reserve food and the length of time they could survive without their host demonstrated that though the mites were starving in the absence of their host, death was not observed at least during the first six hours. This result illustrates that it is possible to run experiments with Varroa mites for at least six hours in the absence of their host. For this reason the calorimetric experiments were limited to 3 to 6 hours, even though no drop in the level of the curve was observed while recording the heat production rate for 10 continuous hours. As can be seen in Fig. 2.3 the mites utilized their energy reserve at a higher rate (8.1% of their wet weight per hour) for the first six hours, with this rate then declining with starvation time. With prolongation of the starvation time they might have run out of reserve food and decreased its utilization rate. With prolongation of the starvation period their reserve food comes to an end and the mites start dying, as seen with the rapidly declining number of survivor mites. Nearly 95% of the starving mites died within the first 36 hours of starvation time, evidently displaying the extent of dependency on their host. The constant proportion of dry weight to fresh weight after different lengths of starvation time indicates that the loss of weight during incubation is not due to evaporational water loss, but because of the utilization of reserve food. The steady proportion of lipid to non-lipid tissue components, regardless of the starvation time, demonstrates that the mites consume lipid and non-lipid reserve food proportionally.

The enhanced weight loss of Varroa mites with ascending temperature indicates that the metabolic rate, and hence reserve food utilization, augmented with increasing temperature. This feature is typical of ectothermic poikilotherms as their metabolic rate and body temperature follow the ambient temperature. The increased consumption of reserve food and the higher metabolic rate at temperatures higher than that in the normal beehive environment demonstrate that the energy and nutritional demand of the mites increase with overheating of the brood nest and hence the mites could cause more damage than at lower temperatures. The loss of wet weight, the heat production, and oxygen consumption rates achieved maximum values at 45 °C and this activity lasted about three hours, after which the rates dropped drastically. The high rates at this temperature indicate that the ambient temperature was intolerable for the mites and that they tried to escape, resulting in a higher metabolic rate and increased utilization of reserve food. The plateau phase in the heat production and oxygen consumption rates between 30°C and 40°C is an indication of the optimum temperature range of Varroa metabolism. Similar patterns obtained with heat production and oxygen consumption rates (Fig. 2.6) point to the fact that it may be possible to use the indirect and cheaper respirometric method in Varroa metabolic investigations if calorimeters are not available.

The phoretic mites have relatively higher heat production rates probably due to the fact that they are fully grown up mites and possess a larger proportion of actively metabolising tissue than reserve food, as compared to the mites from the brood stage, contributing to a higher metabolic rate. A further possible explanation could be that as an adaptation to their way of life, actively attaching themselves to flying bees not to fall down, which needs a larger amount of energy, the phoretic mites may use an efficient metabolising system. Though there was no significant difference in the heat production rate per mite and the weight of individual mites between phoretic and brood mites, the specific heat production rate of phoretic mites was significantly higher than that of brood mites. This could be due to the fact that the heat production rate per phoretic mite was a bit higher and their weight smaller, though not significantly, than that of the brood mites.

Though the eight legged protonymphs and male mites feed on the bee brood and could cause damage (Sammataro et al. 2000) the damage during this phase (the first 3 days of development) was neglected, not to overestimate the impact of the mites in general. Since the largest number of mites per brood cell observed in the present experiments was four in worker brood and six in drone brood, these numbers were used as maxima for the corresponding brood in the computations. In order to compare the heat generation and reserve food consumption of Varroa mites with that of drone and worker brood during ontogenesis the following facts are to be considered: the heat production rate of a drone pupa during the capped developmental stage ranges from 0.7 to 1.9 mW pupa^{-1} with the calculated mean heat production rate and the energy use per day being 1.03 mW pupa^{-1} and 89 J pupa^{-1} respectively. During the whole ontogenetic phase a drone pupa releases 1.34 kJ of energy. The heat production rate of a worker pupa during the capped developmental stage lies between 0.3 and 0.8 mW pupa^{-1} with the calculated mean heat production rate of 0.5 mW. The mean energy use of a worker pupa per day and during the whole ontogenetic phase amount to 43.2 J and 518 J, respectively. The percentage of energy dissipated by Varroa mites as compared to that released by drone pupa ranges from 1.2% (only 3 infesting mites) to 1.6% (6 infesting mites), with the mean value being about 1.3% (4 invading mites). The percentage of energy released by the parasitic mites ranges from 1.2% of that of worker pupa (with only 1 infesting mite) to 2.4% (with 4 infesting mites) with a mean value of 1.4% (with 1.4 mites infesting). As can be seen clearly, the energy demand demonstrated by the heat dissipation rate of a mite, is insignificant even at maximum infestation levels.

The pupal stage of honeybees does not feed and hence depends for the entire ontogenetic process on the reserve food accumulated in the tissue during the larval stage. A non-infested brood achieves a maximum weight during the L6 (182 ± 17 mg for worker brood) and L7 stages (402 ± 19 mg for drone brood). L6 and L7 represent the capped larval stages at the 6th and 7th days of larval development, both at the fifth larval instar. The pupa consumes the reserve food and a freshly hatched worker weighs 105 ± 15 mg and a drone weighs 256 ± 13 mg. Thus, a worker brood and a drone brood utilize 70 mg and 146 mg, respectively, during the entire capped developmental stages. If one compares the amount of reserve food consumed by the developing brood (pupa) during ontogenesis with the amount of hemolymph robbed by Varroa mites subsequently indicating the amount of weight loss by the bee, the mites consume 13% (9.1 mg) (only one infesting mite) to 25% (17.5 mg) (four infesting mites) of the reserve food of the worker brood with a mean value of 15% (10.5 mg) (1.4 infesting mites). Considering the case of drone brood, the infesting Varroa mites rob 16% (23.4 mg) (three infesting mites) to 22% (32.1 mg) (six infesting mites) with a mean value of 19% (27.7 mg) (four infesting mites) of the reserve food of the non-feeding drone pupa.

Energy density analysis of the pupal tissue using bomb calorimetric experiments (Kösece 1998, Contzen et al. 2003) showed that worker and drone pupae possess 26.4 J mg^{-1} and 30.1 J mg^{-1}, respectively. Using these values and the total hemolymph robbed by the mites and comparing it with the total energy dissipated by the them during the capped developmental phase of the pupa, we come up with the result that the mites dissipate only 2.2% of the hemolymph energy they suck from the brood. This result shows that Varroa mites have a very inefficient metabolic machinery and have to feed continuously to fulfil their energy demand. This continuous robbing of the host's reserve food could weaken the colony. Apart from its direct impact, resulting in the formation of underdeveloped and malformed bees, nutritional shortage can also have secondary consequences. These include making the bees vulnerable to viral and bacterial infections, which may otherwise be non-infectious or latent under non-parasitized situations since the immune system of the non-parasitized host could suppress such infections.

The weight loss of workers and drones due to Varroa infestation obtained by the back calculation from the resource utilization of the infesting mites during starvation, and also by the direct weighing of adults immediately before hatching shown in the present results, agrees very well with the results of Schneider and Drescher (1987) obtained by directly weighing the bees immediately after hatching. The authors tabulated their results showing that a worker bee parasitized with 1 to 3 and > 3 mites during brood developmental stage lost 9.6% and 21.6%, respectively, of its unparasitized wet weight. Considering the weight of freshly hatched drones parasitized during the brood stage, Schneider and Drescher (1987) gave a weight loss of 14.1% due to infestation with > 3 mites, in good agreement with the present results of weight loss of 19% when a drone was infested with four mites during its brood developmental stages.

It can be concluded here that the Varroa mites rob a tremendous amount of the reserve food and hemolymph of the brood, contributing to malformations and improper development of wings, abdomen, legs, and the hypopharyngeal glands (Hargasm 1973, De Jong et al. 1982, Schneider and Drescher 1987, Marcangeli et al. 1992). This leads to the development of weak and incapable bees since the reserve food and hemolymph protein are important for ontogenesis and the proper development of the different body parts (Maurizio 1954, Knox et al. 1971). It can not, however, be excluded that factors other than Varroa infestation could play roles in the malformation of bees infested with Varroa mites.

3.0 The Antivarroa Action of Propolis: a Laboratory Assay

3.1 Abstract

Effect of the ethanol extract of propolis (bee glue) against the ectoparasitic mite *Varroa destructor* (Anderson and Trueman) has been investigated and showed narcotic and lethal actions. Length of narcosis and rate of mortality depended on the extraction procedure, concentration of propolis, and contact time. Propolis extracted with 70% ethanol was found to be highly toxic, a 10% (w/v) propolis resulting in 100% mortality with a brief contact time of 5 s. In addition, the effect of propolis on the metabolic rate of Varroa mites has been investigated calorimetrically. Even sublethal propolis concentrations without varroacidal effects, and of only short lasting narcotic effects, resulted in a significant reduction of the specific heat production rate, indicating that the mites are weakened.

3.2 Introduction

The threat of honeybee infestation by *Varroa destructor* (Anderson and Trueman) forces beekeepers in many parts of the world to treat their colonies with acaricides, which, however, are associated with drawbacks. The most serious drawbacks are the build up of residues in bee products (Kubik et al. 1995, Wallner 1995, Stürz and Wallner 1997, Bogdanov et al. 1998, Wallner 1999) and the emergence of resistant mite strains. Varroa strains resistant to the different types of acaricides in use today have been reported from different parts of the world. Resistance of mites to fluvalinate and flumethrin have been reported in Europe and the United States (Colombo et al. 1993, Eichen 1995, Lodesani et al. 1995, Milani 1995, Baxter et al. 1998), to coumaphos in Italy and Switzerland (Milani and Della Vedova 1996), to amitraz in the United States (Elzen et al. 2000), and to bromopropylate and chlordimeform in Europe (Ritter and Roth 1988).

The above mentioned problems associated with the use of acaricides provide considerable incentives to develop new treatment strategies and screening for potential acaricides that minimize these problems. Natural products having different components with various modes of action might provide effective solutions to the problems of varroosis (Mutinelli et al. 1997, Imdorf et al. 1999). One of such natural products is propolis (bee glue), a complex mixture of several compounds collected by honeybees from plants and used in the construction and protection of the beehive (Ghisalberti 1979).

Literature on the acaricidal or insecticidal action of propolis is very limited. It has been assumed that components of nectar, pollen, and propolis may adversely affect the development of *V. destructor* in the hive of some bee populations rather than emergence of natural resistance (Amrin et al. 1996, http://www.wvu.edu/agexten/varroa.htm). It has been suggested that some flavonoid components of propolis have insecticidal or at least insectistatic (inhibition of insect larval development) effects (König and Dustmann 1988). Even though the anaesthetic and lethal actions of propolis against *V. destructor* have been briefly mentioned in the literature (Schkurat and Poprawko 1980), its potential acaricidal use is not yet investigated.

3.3 Materials and Methods

3.3.1 Propolis extraction and preparation

Propolis samples used in the present experiments were obtained by scrapping off the frames from beehives in the garden of the Institute of Zoology, Free University of Berlin, Germany. Pre-weighed and frozen samples were homogenised using a coffee mill (type MZ Moulinex, France). The homogenate powder was then extracted in 70% or 40% ethanol. The extraction in 40% ethanol was intended to procure components to be used in less concentrated ethanol solution to minimize the effect of ethanol on the experimental organisms. For effective extraction the propolis powder was suspended in the corresponding ethanol solution in a ratio of 1:9 (w/v) (Strehl et al. 1994). The suspension was extracted in a rotary evaporator (Rotationsverdampfer W-micro, Heidolf, Mannheim, Germany) at 60 °C for 2 h. The suspension was then cooled at room temperature for ca. 1 h and then suction filtered. The filtrate was dried in an incubator at 40 °C to weight constancy, which was achieved in two weeks time. The yield of extraction was 58% (w/w) for the extraction in 70% ethanol and 19% (w/w) in 40% ethanol.

The 70% ethanol extract was used in 55% ethanol (**solution B** hereafter) in the bioassay in order to reduce the effect of strong ethanol solution on the experimental organisms. The little amount of precipitation observed while suspending solution B was brought into solution by agitation. The 40% ethanol extract was used in the same ethanol concentration (**solution A** hereafter) for the bioassay. The concentrations used in the bioassay were 5%, 7.5%, 10%, 15%, 20% (w/v) solution A and 0.5%, 1%, 2%, 5%, 7.5%, 10% (w/v) solution B.

As the presence of acaricide residues in the propolis sample, due to previous treatments of the colony, may introduce artefacts, acaricide residue analysis was carried out to assure the propolis quality. The residue analysis of the propolis sample was done at the Landesanstalt für Bienenkunde der Universität Hohenheim, Stuttgart, Germany.

3.3.2 Mite collection

Mites were collected from infested colonies, treated only at the beginning of autumn of the preceding year with formic acid, in the garden of the Institute of Zoology, Free University of Berlin, Germany. The experiments were conducted in summer 2000. Adult Varroa females were collected from capped healthy drone brood by opening and inspecting individual cells. During the collection process mites were kept in a Petridish on bee larvae or pupae in order to avoid starvation. Newly moulted adult mites, identified by their pale colour, relatively smaller size, and weak locomotion, were excluded from the experiment, since they may have a different response as hardening of the cuticle is still in progress. Mites which seemed weak and abnormal were discarded.

3.3.3 Bioassay

Treatment of the mites was achieved by applying 250 µl of a given concentration of propolis on a 3 cm x 3 cm tissue paper (Kimwipes™ Lite 200, Kimberly – Clark™) in a Petridish and by immediately placing six mites per experiment on the wetted tissue paper. In order to observe the effect of contact time of propolis on the activity of *V. destructor* the following treatment times were used: 5, 10, 20, 30, 40, 60, 75, 90 s for the treatment with solution B and 20, 40, 60, 90 s for the treatment with solution A. The treatment was stopped after the allocated time by removing the mites with the tissue paper from the Petridish, and immediately placing them on a pad of paper towel for 1 min to blot the excess fluid on the surfaces. They were then transferred to a clean Petridish, and their activity was observed under a dissecting lens. Their activity was checked every five minutes for the first hour, every 10 minutes for the next one hour, and every 30 minutes for the next two hours. All treatments were carried out at room temperature (25 °C) and the treated mites were incubated at 35 ± 0.5 °C. Control experiments for each experimental group were conducted by treating the mites for the corresponding time with 40% or 55% ethanol solution and also distilled water.

An individual mite was considered inactivated if it showed no leg movement or movement of any body part when gently prodded with a probe. If it showed movement it was counted as alive, irrespectively of whether it was partially paralysed or normal. If the inactivity lasted more than four hours after the treatment time the mites were considered dead; further incubation did not show any activity change. Each treatment was repeated five times and the mean ± S.D. values were used in the presentation of results.

3.3.4 Calorimetric experiments

The bioassay method mentioned above enables us to assess the action of propolis only by counting the number of active or inactivated (dead) individuals, but not the extent of the effect on the surviving and weakened individuals. For this reason calorimetric experiments were conducted to observe to what extent a certain sublethal propolis dose affects the metabolic rate of the mites. The calorimeter used was a Biocalorimeter, B.C.P-600 (MV Messgeräte Vertrieb, München, Germany) with a sensitivity of 44.73 μV mW^{-1} and a vessel volume of 12 cm^3.

In order to compare the heat production rate before and after treatment, 20 to 25 untreated mites per experiment were put in the calorimetric vessel and their heat production rate was recorded for 2 h. Recording was then stopped, and the mites were removed from the calorimeter and treated with propolis. The treatment lasted 30 s with solution B and 60 s with solution A. The treated mites were put back into the calorimetric vessel and their heat production rate was recorded for 4 to 5 h. Each experiment was repeated 5 times and the mean ± S.D. values were used in the presentation of results.

3.4 Results

The acaricide residue analysis of propolis, with a detection limit of 1 mg kg^{-1}, showed that the propolis sample was free of any acaricide contaminant.

The control treatments of both solution A and solution B, i.e. with only 40% and 55% ethanol, respectively, have shown narcotic effects for a short period after treatment. In both cases this effect lasted < 5min i.e., 100% of the mites recovered within this time interval. The proportion of narcotised mites just after the treatment (zero observation time) ranged from 46.7 ± 7.5% to 53.3 ± 11.8 % for control of solution A at 20 s and 90 s contact times, respectively. In case of the control of solution B, 53.3 ± 13.9% to 100% of the mites were narcotised at treatment times of ≤ 10 s, and ≥ 60 s, respectively with the rest lying between these values. Even though the control mites of solution A were narcotised shortly after treatment, their metabolic activity was not significantly affected after recovery ($P > 0.05$); being13.1 ± 1.1 (μW mg^{-1}) and 13.0 ± 1.4 (μW mg^{-1}) before and after treatment, respectively

(Fig. 3.5 a). The control treatments of solution B, however, displayed a significant effect (P < 0.05) on the metabolic activity of the mites resulting in a drop of the specific heat production rate from 13.6 ±1.8 (μW mg^{-1}) to 11.8 ± 2.2 (μW mg^{-1}) by 13% after treatment (Fig. 3.5 b). Contact with water had no effect at all.

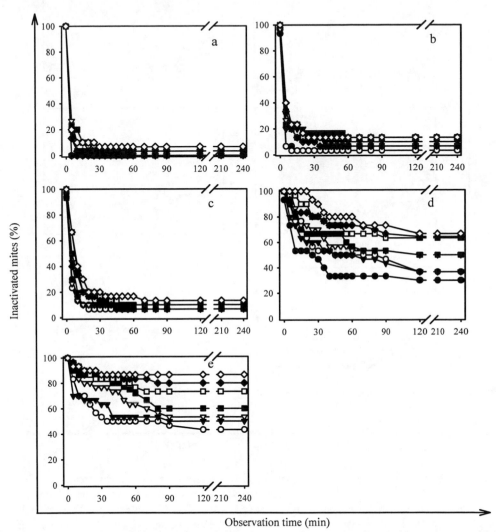

Fig. 3.1 Effect of contact time on the activity of *Varroa destructor* mites under treatment with 0.5% (a), 1% (b), 2% (c), 5% (d), and 7.5% (e) w/v propolis in 55% ethanol (solution B). Six mites per experiment, n = 5. ● - 5s, o - 10s, ▼ - 20s, ∇ - 30s ■ - 40s - 60s, ◆ - 75s, ◇ - 90s contact time. Even though the control treatments displayed narcosis of 53.3 to 100 % of the mites, directly dependent on the treatment time, narcosis lasted less than 5 min in all cases and, hence, was not incorporated in the graph.

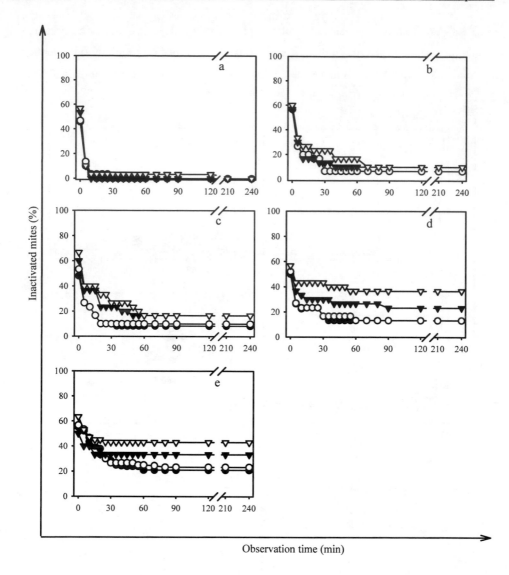

Fig. 3.2 Effect of contact time on the activity of *Varroa destructor* mites under treatment with 5% (a), 7.5% (b), 10% (c), 15% (d), and 20% (e) w/v propolis in 40% ethanol (solution A). Six mites per experiment, n = 5. ● - 20s, O - 40s, ▼ - 60s, ▽ - 90s contact time. Even though the control treatments displayed narcosis of 46% to 53 % of the mites, directly dependent on the treatment time, it lasted less than 5 min in all cases and, hence, was not incorporated in the graph.

Treatment of *V. destructor* mites with various propolis concentrations of solution B at different contact times showed that 100% narcosis was achieved for some minutes immediately after treatment, regardless of propolis concentration and contact time. Further observation of activity of the mites displayed that with lower concentration of solution B

(0.5%, 1% and 2%) narcosis lasted shorter (Fig. 3.1 a, b, c) and most mites recovered from narcosis in the first 5 to 15 minutes after treatment. Even if recovery was observed after this time it was very slow. Narcosis with higher concentrations of solution B (5%, 7.5%) lasted longer and most mites that could recover, recovered within the first 90 minutes after treatment (Fig. 3.1 d, e). Very few mites were observed recovering two hours after treatment. Treatment of *V. destructor* with solution A resulted in an initial narcosis of 46.7% to 53.3% (Fig. 3.2 a to e) which is lower in comparison with that of solution B. Most mites treated with lower concentration of solution A (5%) recovered within the first 10 min (Fig. 3.2 a), regardless of the contact time. Treatment time played a role in case of treatments with higher concentrations of propolis for both solutions A and B (Fig. 3.1 d, e, and Fig. 3.2 b, c, d, e). Treatment of mites with 10% (w/v) propolis in 55% ethanol (10% solution B) resulted in 100% mortality, regardless of the treatment time, indicating its high toxicity with the slightest contact time, as low as 5 s. As there was no activity of mites observed after treatment with 10% solution B it was not necessary to display this in Fig. 3.1.

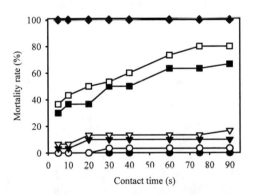

Fig. 3.3 Influence of contact time on the mortality rate of *Varroa destructor* under treatment with different concentrations (% w/v) of solution B (in 55% ethanol). Six mites per experiment, n = 5, percentages of the mean values are presented here. ●- control, ○- 0.5%, ▼- 1%, ▽- 2%, ■- 5%, □- 7.5%, ◆- 10% propolis. 4 h after treatment.

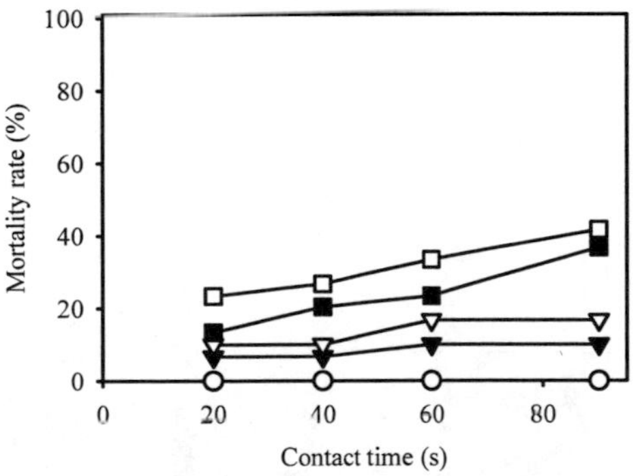

Fig. 3.4 Influence of contact time on the mortality rate of *Varroa destructor* under treatment with different concentrations (% w/v) of solution A (in 40% ethanol). Six mites per experiment, n = 5, percentages of the mean values are presented here. ●- control, ○- 5%, ▼- 7.5%, ▽- 10%, ■- 15%, □- 20%. The control experiment and 5% had no lethal effects, values overlapping at the x-axis. 4 h after treatment.

Even though some mites were observed recovering from narcosis, others could not recover at all. Mites that did not recover in the first 4 h after treatment were dead. This was confirmed by incubating the treated mites for a total of 8 h. The effect of different propolis concentrations and contact times on the mortality rate of *V. destructor* is demonstrated in Fig. 3.3 and Fig. 3.4. The varroacidal action of propolis increased with increasing concentration and contact time except for lower concentrations of both solutions (Fig. 3.3, and Fig 3.4). Treatment of mites with 10% (w/v) solution B resulted in 100% mortality regardless of the treatment time, indicating its high toxicity with the slightest contact time. Treatment of mites with solution A, even with a concentration of 20% (w/v), resulted in a mortality rate less than 50% (Fig. 3.4).

Comparison of the specific heat production rates of Varroa mites before and after treatment with different concentrations of solutions A and B showed that even those concentrations that did not have a considerable effect on the mortality rate of mites dropped the specific heat production rate significantly (paired sample t-test, $p < 0.05$, to $p < 0.001$, Fig. 3.5 a and b). The reduction in the specific heat production rate due to treatment with propolis grew with increasing concentrations (Fig. 3.5 a and b insets), except at higher concentrations of propolis in 55% ethanol (solution B), where a saturation effect was observed (Fig. 3.5 b inset).

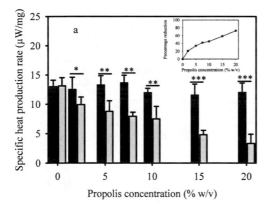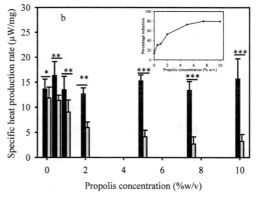

Fig. 3.5 Effect of different concentrations (% w/v) of propolis in 40% ethanol - solution A (a) and in 55% ethanol - solution B (b) on the mean specific heat production rate (mean ± S.D.) of *Varroa destructor* mites. 20 to 25 mites per experiment, n = 5. The inset in each graph, extracted from the corresponding graph, is a curve of percentage reduction of the mean specific heat production rate versus propolis concentration of treatment. ■ - before treatment, ☐ - after treatment. The values at zero concentration are treatments with the corresponding ethanol solutions (controls). Significance levels of $P < 0.05$ *, $P < 0.01$**, $P < 0.001$*** (Paired sample t-test).

A feature common to all power-time (*p-t*) curves obtained after treatment with propolis, in addition to the reduction in the heat production rates, was the loss of the typical structures and subsequently smoothing of the curves.

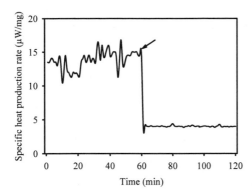

Fig. 3.6 Effect of treatment of *Varroa destructor* mites with 5% (w/v) propolis in 55% ethanol (solution B) on the structure and the level of the specific heat production rate (μW mg^{-1}) in a typical calorimetric experiment with 25 mites. After treatment with propolis, a period of 45 min (omitted in the graph) was required before starting to record the heat production rate again. The arrow indicates point of treatment.

The difference between the maximum and minimum points on the curves diminished progressively after treatment with increasing concentration of propolis. With treatments at higher concentrations of propolis the curves became nearly smooth, and lay at lower levels. Fig. 3.6 displays a typical example of a *p-t* curve whose structure was highly affected by the treatment with propolis.

3.5. Discussion

Treatment of mites with propolis causes narcosis and death. The length of narcosis and perhaps the subsequent death of mites after treatment with propolis concentration above a certain threshold depend on propolis concentration, solvent of extraction and length of contact time. This does not, however, hold true at lower concentrations, like in 0.5%, 1% and 2% solution B, or 5% of solution A. This is due to the fact that even if the length of contact time is prolonged, the concentration of bioactive components that penetrate into the mites' body is too low to cause considerable harm and, therefore, remains at a sublethal concentration. The length of contact time plays a role due to the fact that with the increase in the length of contact time, the amount of bioactive components penetrating into the mites' body increases, raising its concentration in the animal tissue, and achieving lethal doses. The slightest contact of Varroa mites with 10% solution B, regardless of the contact time, resulted in 100% mortality indicating that it is highly toxic. The narcotic effect of propolis on different animals has already been mentioned in the literature (Prokopovich et al. 1956, Prokopovich 1957). As seen from the present experimental results, treatment of mites with solution B was more effective than the corresponding treatment with solution A. The most plausible explanation for these differences is that solution B was extracted in 70% ethanol whereas solution A was extracted in 40% ethanol. The extraction of propolis in 70% ethanol enables the procurement of most of the biologically active hydrophobic components, which could not be extracted in 40% ethanol. This means that solution B was qualitatively and/or quantitatively superior to solution A.

Even though the control experiments of solutions A and B showed differences in the percentage of narcotised mites shortly after treatment, and in the reduction of heat production rates, there was no mite mortality observed in both cases. Unlike the controls, the treatments with different concentrations of the two propolis solutions showed considerable differences not only in their effect on narcosis and heat production rate, but also on the mortality rate of mites. This indicates that it is not the concentration of the alcohol, rather the difference in the ingredients of the two solutions of propolis that is responsible for the observed difference in mite mortality after the treatments with the two solutions.

Even though the treatment of mites with the different concentrations of solution A and weak concentrations of solution B (0.5%, 1%, and 2%) displayed only little varroacidal effects (Fig 3.3 and 3.4), it affected the specific heat production rate significantly (Fig. 3.5 a and b). As an example, treatment with 20% solution A with a contact time of 60 s killed only 33% of the mites, whereas the specific heat production rate was reduced by 75%. This is an indication that even the non-lethal or feebly lethal doses of propolis strongly emasculate the mites, and that the mites could not perform their locomotion nor were they able to move their body parts. The control experiments did not have lethal effects at all, but they affected the heat production rate. This fact suggests that monitoring the metabolic/heat production rates using calorimetric methods is more sensitive than observation of their locomotory activity, in order to judge the effect of propolis on mites. A limitation of the calorimetric method in the investigation of the action of propolis on Varroa mites was that the heat production rate during the recovery process from narcosis was difficult to follow. This difficulty arose due to the fact that the time required for the thermal equilibration of the calorimeter after opening it, to place the samples, was long (30 to 45 min). During this thermal equilibration time most of the mites have already recovered and attained a certain steady state metabolic rate. Had it not been for this limitation, the calorimetric method would have helped us to observe the heat production rate while the mites were narcotised, and in the gradual recovery process from narcosis

A feature common to almost all *p-t* curves obtained after the treatment with higher concentrations of propolis was the smoothness of the curve, and its position at lower level than before treatment. The highly structured *p-t* curves, with clearly distinct maximum and minimum signals were smoothed due to the treatment (Fig. 3.6), the extent of smoothing of the curve increasing with rising concentrations of propolis. This phenomenon may be explained as follows: all the mites may rest at a time, resulting in a minimum specific heat production rate, and all of them become very active simultaneously resulting in a maximum heat production rate. As an adaptation to their way of life the mites produce faeces (guanine) which is dry (Sammataro et al. 2000). Thus the deposition of faeces in the calorimetric vessel may not cause a considerable evaporational heat loss, and hence the troughs are mainly due to the simultaneous resting of mites. This feature of rest and simultaneous activity of the mites was observed outside the calorimeter, after placing them in a small glass vessel. Observation of their activity demonstrated that a mite rests for some time and starts moving again chaotically, disturbing the mites in the vicinity that respond in the same fashion. This process of chaotic activity lasts for some minutes and all the mites in the vessel may become very active, resulting in the maximum specific heat production rate. The phenomenon of rest and simultaneous activity is responsible for the structuring of the *p-t* curve. The smoothness of the *p-t* curve after treatment with propolis results due to the fact that the mites are highly weakened, and unable to perform their usual movement. They were able to move feebly or perform their resting and weakened metabolic activity, resulting in a *p-t* curve which is smooth.

The present experimental results showed that *V. destructor* mites are highly sensitive to propolis solution *in vitro*. It has been postulated (König and Dustmann 1988) that bees must be getting some benefit from the use of propolis in the beehive; otherwise they would not waste time and energy in collecting it. Thus, it is possible that propolis could affect Varroa mites in the beehive to a certain extent. The varroacidal action of propolis seems to be paradoxical, since propolis and Varroa mites are normally found in the beehive, the mites walking on thin propolis layers throughout the hive. The most plausible explanation as to why propolis does not kill Varroa mites in the beehive is that propolis is insoluble in the beehive interior. This is because most of the components of propolis are water insoluble. The water soluble components of propolis that could potentially affect mites in the beehive *in situ* comprise about 2.5% to 6.5%, based on the origin of propolis (Neunaber 1995). As seen in the case of efficacy (both narcosis and mortality) of solution A, where most of the water soluble and some water insoluble components are extracted, a higher concentration of propolis was needed to observe a remarkable varroacidal effect. This fact displays that even if some of the components of propolis are dissolved in the high humidity in the hive's interior, their concentration is too weak to remarkably affect/kill the mites.

If propolis is to be recommended for use as a varroacidal agent it may minimize the contamination of hive products by synthetic acaricides. Except for those people that are allergic to propolis, its presence in hive products may not be considered as a serious contaminant. Indeed propolis is already being used in some countries as an additive of cosmetics and in medicine.

In order to reduce the amount of unnecessary chemicals of propolis in hive products the active varroacidal components of propolis may be fractionated and used separately. In addition to this, the synergistic action of propolis with essential oils, already being used as varroacides, may have to be investigated. If propolis is effective in the field experiment, and if it has no negative effect on the bees themselves, it may minimize the cost of beekeeping.

Propolis from different geographic origins differ from each other in their chemical makeup and, hence, probably in their Varroa narcotizing and varroacidal actions. Since the antivarroa action of propolis could be affected by temperature, the next two chapters will concentrate on the investigation of the antivarroa action of propolis from different geographic locations, at different experimental temperatures.

4.0 Microcalorimetric and Respirometric Investigations of the Effect of Temperature on the Antivarroa Action of Propolis

4.1 Abstract

The antivarroa action of propolis and its synergism with temperature was investigated calorimetrically and respirometerically using female *Varroa destructor* (Anderson and Trueman) mites from adult workers, worker and drone brood.

The treatment of Varroa mites with 4% propolis in ethanol affected the metabolic activity of the mites with the effect directly related to the temperature of treatment. The mites collected from worker and drone brood reacted similarly to propolis treatment at different temperatures. In contrast to that, the mites from adult workers (phoretic mites) responded differently; the treatment with 4% propolis at 40 °C resulted in 100% mortality of mites from adult workers, but only reduced the heat production rate of mites from worker and drone brood by 68% and 60%, respectively. The changes in heat production and oxygen consumption rates, as a function of temperature, showed similar patterns before as well as after treatment with 4% propolis.

Exposure of mites to 45 °C agitated them, as witnessed by the elevated heat production rates, 23.5 ± 2.5 μW mg^{-1} compared to 14.4 ± 1.0 μW mg^{-1} at 35 °C. After treatment with propolis at 45 °C, all mites died, regardless of their origin, indicating that the simultaneous use of varroacides and high temperature treatment for a short period of time could be more effective than the prolonged use of either method.

4.2 Introduction

Varroa destructor (Anderson and Trueman) is a serious ectoparasitic parasite of the western honeybee *Apis mellifera* L., infesting both feral and managed colonies. It has caused the destruction of numerous colonies. This led to the reduction in the number of beekeepers and the harvest of honey and other bee products in different parts of the world, in the last three decades (Peng ct al. 1987, Delfinado-Baker 1988, Kovac and Crailsheim 1988, Matheson 1994, Kraus and Page 1995, De Jong 1997, Finley et al. 1997). To save their colonies from obliteration beekeepers are using acaricides as short-term solutions. The use of chemical acaricides is not, however, free from drawbacks: accumulation of residues in bee products (Kubik et al. 1995, Wallner 1995, Stürz and Wallner 1997, Bogdanov et al. 1998, Wallner 1999), hazards to the bees and/or the beekeeper (Ellis Jr. 2001), and the emergence of acaricide resistant mites (Milani 1994, Lodesani et al. 1995)

There is an ever increasing number of reports concerning the resistance of *Varroa destructor* against various acaricides in different parts of the world. The problems associated with the use of acaricides provide incentives to bee researchers and beekeepers to search for better acaricides or methods to control Varroa mites. Different methods, like the biotechnical mite control (Rosenkranz 1987, Maul et al. 1988, Fries and Hansen 1993, Engels 1994, Calis et al. 1998), heat treatment of infested brood using a Mitezapper (Huang 2001) and infested adult workers (Tabor and Ambrose 2001), a combination of biotechnical control and acaricide treatment (Fries 1991, Calis et al. 1998), a combination of biotechnical control and heat treatment (Rosenkranz 1987, Engels 1994) have been shown to be effective in the control of Varroa mites. Of particular interest in the search for new acaricides are compounds that are natural in origin. Botanical extracts and essential oils have exhibited some efficacy as means to control Varroa mites (Imdorf et al. 1999, Ellis Jr. 2001). Many plants produce essential oils or other chemicals that are used as natural pesticides to ward off insect herbivores or prevent infection of wounds. One of these groups of chemicals produced by plants is propolis.

A laboratory assay of the antivarroa actions of propolis in Chapter 3 displayed that it possesses both narcotic and lethal actions with 10% propolis killing 100% of Varroa mites, even with a very short contact time of 5 s. Lower concentrations display varying degrees of suppression of the metabolic rate of the mites.

In this chapter it will be demonstrated, by calorimetric and respirometric methods, if the varroacidal action of propolis can be augmented by a simultaneous exposure of the mites to higher or lower temperature extremes.

4.3 Materials and Methods

4.3.1 Animal material

Infested honeybee colonies of the bee race *Apis mellifera carnica* from the research beehives of the Institute of Zoology, Free University of Berlin, Germany were used as sources of *Varroa destructor* mites for the present experiments. The experiments were conducted in summer 2001. At the beginning of autumn of the previous year the experimental colonies were treated once with formic acid to reduce the infestation level and eventual annihilation of the colony by Varroa mites.

Female *Varroa destructor* mites were collected from adult workers, worker brood, and drone brood. The collection of mites from the brood stage was carried out at room temperature by opening and inspecting healthy brood. During the collection process mites were kept in a Petridish on the corresponding bee pupae in order to avoid starvation. Newly moulted adult mites were excluded from the experiment since they might be possible sources of error given that the development of the cuticle could still be in progress. Mites that seemed weak and abnormal were discarded. Collection of mites from adult workers was done by very cautiously dislodging them from the surface of the bees with the help of a blunt needle.

4.3.2 Calorimetric experiments

The calorimetric experiments were performed using two isoperibolic calorimeters: (i) a Biocalorimeter B.C.P. (Messgeräte Vertrieb, München, Germany) with a sensitivity of 44.73 μV mW^{-1} and a vessel volume of 12 cm^3 and (ii) a THERMANALYSE calorimeter (Messgeräte Vertrieb, München, Germany) with a sensitivity of 40.65 μV mW^{-1} and a vessel volume of 15 cm^3. The calorimetric vessels are big enough to provide adequate oxygen for the entire experimental period. The calorimetric experiments were conducted at temperatures of 25, 30, 35, 40, 45, and 50 °C with mites from adult workers, worker brood, and drone brood. 20 to 30 mites were weighed before each experiment using an analytical balance (Sauter, Ebingen, Germany) of 0.1 mg sensitivity, transferred into the calorimeter, and the heat production rate was recorded for 2 to 3 h. In case of experiments with 45 °C, the heat production rate before treatment was recorded only for 45 min, since the mites started dying within 90 to 180 min after exposure. Recording of heat production rate was stopped after the pre-selected experimental time and mites were removed from the calorimeter, and weighed again immediately to find out the change in weight. The rate of utilization of reserve food under starving conditions and, thus, the change of weight help to illustrate the amount of hemolymph the mites could utilize from their host to maintain their weight and physiological status under natural and non-starving conditions. After weighing the mites were immediately treated with a solution of 4% propolis in 55% ethanol as described below (4.3.3). Having blotted the excess fluid from their surface the mites were weighed again to obtain the after-treatment initial weight, and they were put back into the calorimeter and their heat production rate was recorded further for 3 to 5 h. The mites were weighed at the end of the calorimetric experiment to determine the rate of weight change. The hypothetical weight loss of a mite per day was extrapolated from the weight loss during the experimental period. In this regard the rate of weight change was presented as wet weight loss per mite per day before and after treatment with 4% propolis at the different experimental temperatures.

4.3.3 Treatment of mites with propolis

Since the goal of these experiments was to observe the effect of temperature on the antivarroa action of a sublethal dose of propolis, a 4% propolis in 55% ethanol was used. It made no sense to apply a lethal dose of propolis because calorimetry after treatment with such doses is irrelevant. The propolis used for these experiments was obtained from Holeta Bee Research Center, Ethiopia. It was extracted in a rotational evaporator (Rotationsverdampfer W-micro, Mannheim, Germany) for 2 h in 70% ethanol. The dried extract was dissolved in 55% ethanol for further use.

In preparation for treatment the mites were placed in a clean Petridish, on top of a 3 cm x 3 cm tissue paper (Kimwipes™ Lite 200, Kimberly – Clark™). Mites were treated for 30 s after applying 250 µL of the 4% propolis solution on the tissue paper, not directly on the mites. After the allocated time the treatment was ended by removing the mites from the Petridish, and placing them on a pad of paper towel for 1 min to blot the excess fluid on their surfaces. Blotting of the excess fluid from the mites' surface after the conclusion of the treatment was important, since it otherwise would have interfered with the calorimetric signal due to evaporational heat loss. The treated and blotted mites were weighed again, placed back into the calorimetric vessel, and their heat production rate was recorded. Control experiments for each experimental group were carried out by treating the mites with 55% ethanol as well as distilled water.

4.3.4 Respirometric experiments

The effect of temperature and propolis treatment on the oxygen consumption rate of *Varroa destructor* mites from drone brood was investigated at 25, 30, 35, 40, 45, and 50 °C using manometric methods. The respiration experiments were conducted using Warburg vessels of about 12 cm^3 volume and 50 to 60 mites per experiment. CO_2 produced during respiration was absorbed by a 4% KOH solution. In order to avoid access of the mites to the KOH solution, the opening to the side arm was fitted with a very thin layer (1mm thick) of porous spongy material with a pore size of ca. 0.8 mm, which allows air to enter but not the mites. Recording the oxygen consumption rate was started after a temperature equilibration time of 30 min, and further recording was made in intervals of 30 min, for 2 h before treatment. The respirometric experimental time for each temperature set-up was equal to the calorimetric experimental times. Each measurement was conducted 5 times, but 9 times in case of measurements with 45 °C since the experimental time of the latter temperature set-up was short compared to that of the others due to mite death with prolonged experimental period.

Finally comparison of the effect of propolis at different experimental temperatures on the metabolic rate of mites from the various developmental stages will be made to see if the mites have different responses.

4.3.5 Statistical analysis

Results were presented as mean ± S.D. values. The level of difference in the heat production rate, oxygen consumption rate, and weight loss rate at the different experimental temperatures, before and after treatment with propolis were determined using either the student's t-test, or paired sample t-test, or 1-way ANOVA and Tukey's HSD post hoc test with a critical value of $\alpha = 0.05$.

4.4 Results

The mites obtained from adult workers, worker brood, and drone brood died immediately after exposure to 50 °C in both calorimetric and respirometric experiments. Hence the results at this temperature set-up are missing in most graphs since the rates are nil.

Mites collected from worker and drone brood showed comparable specific heat production rates at different experimental temperatures, except at 45 °C where the mites from worker brood exhibited a significantly higher heat production rate (students t-test, p = 0.03). At an experimental temperature of 25 °C, the mites collected from adult workers displayed a significantly higher specific heat production rate of 8.4 ± 1.4 $\mu W\ mg^{-1}$, compared to 5.0 ± 0.4 $\mu W\ mg^{-1}$ (p = 0.025) and 6.1 ± 1.2 $\mu W\ mg^{-1}$ (p = 0.033), for mites from worker brood and drone brood, respectively (1-way ANOVA, p = 0.019, and Tukey's HSD post hoc test) (Fig. 4.1). There was no significant difference (p = 0.16) between the heat production rates of mites from worker and drone brood. The mites from adult workers also had a significantly higher specific heat production rate of 17.4 ± 1.1 $\mu W\ mg^{-1}$ at 35 °C, compared to those from worker brood with 14.3 ± 0.9 $\mu W\ mg^{-1}$ (p = 0.041) and drone brood with 14.6 ± 1.0 $\mu W\ mg^{-1}$ (p = 0.036) (1-way ANOVA, p = 0.025, and Tukey's HSD post hoc test). Mites from worker and drone brood showed no significantly different heat production rates (p = 0.35) at 35 °C too. There was no significant differences in the heat production rates of the different mites at 30 °C and 40 °C (p = 0.08, and p = 0.12, respectively, 1-way ANOVA). Regardless of where the mites were obtained from, the heat production rates increased with raising calorimetric temperature, achieving constant rates between 35 °C and 40 °C. With the shift of temperature from 40 to 45 °C the heat production rate increased drastically from 14.6 to 23.5 $\mu W\ mg^{-1}$, in case of mites from worker brood, as an example. The high heat production rates at this elevated temperature (45 °C) lasted for a short period of time, only 90 to 180 min. After this time interval the *p-t* curves declined, due to the death of mites; death ensued faster in case of mites from adult workers than those from worker and drone brood.

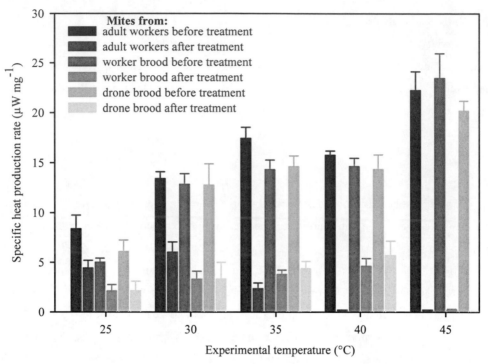

Fig. 4.1 The effect of temperature on the specific heat production rate of *Varroa destructor* mites before and after treatment with 4% propolis. Twenty to 30 mites per experiment, n = 5, mean ± s.d. The treatment with 55% ethanol (control) reduced the heat production rate by 5 % to 9% regardless of temperature and origin of mites.

Treatment of mites with 4% propolis resulted in a reduction of the heat production rate. The extent of reduction increased with the experimental temperature, especially in case of mites obtained from adult workers (Fig. 4.2). Mites from adult workers were all dead after treatment with 4% propolis at 40 °C, whereas those from worker and drone brood showed a reduction in the heat production rate by 68% and 60%, respectively. A mortality of 100% was achieved after treatment with 4% propolis at 45 °C, regardless of the origin of mites, the heat production rate dropping to the base line (Fig. 4.1 and 4.2). Varroa mites from worker and drone brood showed nearly similar responses to the treatment with propolis at different experimental temperatures, whereas mites from adult workers had a different response (Fig. 4.2). The control experiments, treatments with 55% ethanol rendered a reduction of the specific heat production rate by 8% to 11% regardless of the temperature of treatment.

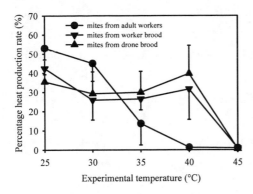

Fig. 4.2 Effect of temperature on the percentage residual heat production rate (p after treatment / p before treatment x 100) of *Varroa destructor* mites after treatment with 4% propolis. Twenty to 30 mites per experiment, n = 5, mean ± s.d. After treatment with 55% ethanol (control) the residual heat production rate lay between 91 % and 95% regardless of the experimental temperature and origin of mites.

Heat production and oxygen consumption rates of Varroa mites from drone brood behaved similarly before and after treatment with 4% propolis at different experimental temperatures (Fig. 4.3). With the increase of temperature by 10 K from 25 to 35 °C the specific heat production rate (Q_{10}) before treatment increased by a factor of 2.4; the corresponding oxygen consumption rate increased by a factor of 2.3. Considering the temperature interval between 30 and 40 °C where the metabolic rate is nearly constant, the heat production rate increased by a factor of only 1.1, and the oxygen consumption rate grew by a factor of 1.2. After treatment with 4% propolis the changes in the rates of heat production and oxygen consumption showed different patterns than before treatment. With the temperature increase from 25 to 35 °C the heat production rate changed by a factor of 2.0 whereas the oxygen consumption rate by 1.5. The Q_{10} value after treatment with 4% propolis for the shift from 30 to 40 °C amounted to 1.7, and the oxygen consumption changed by a factor of 1.5 (Table 4.1). The treatment with 55% ethanol (control experiment) reduced the oxygen consumption rate by 6 to 10% independent of the experimental temperature. The other control experiments showed no effect.

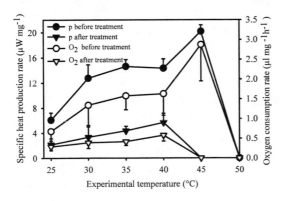

Fig. 4.3 Effect of temperature on the specific heat production rate (p) and oxygen (O_2) consumption rate of *Varroa destructor* mites from drone brood before and after treatment with 4% propolis. 20 to 30, and 50 to 60 mites per experiment, for the calorimetric and respirometric experiments, respectively. n = 5 (but n = 9 for the respirometric measurements at 45°C), mean ± s.d. The oxygen consumption rate is reduced by 5% to 11% and the heat production rate by 5% to 9% in the control group

Utilization of own reserve food by Varroa mites, displayed by the loss of wet weight during starvation, was highly affected by the experimental temperature. The mites collected from adult workers lost a higher proportion of their body weight per starvation hour than those from drone and worker brood. A mite from an adult worker may utilize 1.5 fold of its own weight per day at 25 °C, whereas those from worker and drone brood could utilize 1.2 and 1.1 fold of their own weight, respectively, at this temperature. The rate of wet weight loss by mites from adult workers was significantly different from the other two, which do not display significant difference among each other (1-way ANOVA, Tukey's HSD post hoc test $\alpha = 0.05$).

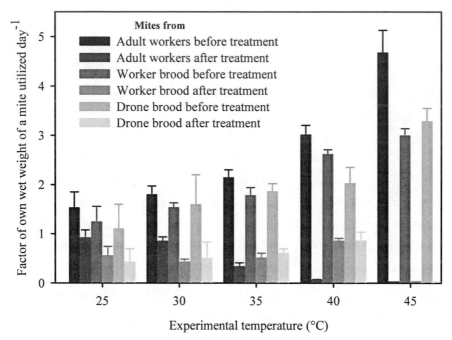

Fig. 4.4 Experimentally determined weight loss of the mite *Varroa destructor*, extrapolated to a hypothetical value per day and presented as a factor of the initial wet weight, under starvation and different experimental temperature conditions, before and after treatment with 4% propolis. n = 5, 25 to 30 mites per experiment. Treatment with 55% ethanol (control) reduced the weight loss rate by 7 % to 12% regardless of mites' origin and temperature.

At 45 °C mites from adult workers, worker brood or drone brood could theoretically utilize 4.7, 3.0 or 3.3 fold, respectively, of their own weight per day (Fig. 4.4). Percentage of wet weight reduction after treatment with 4% propolis declined drastically with increasing temperature, especially in case of mites from adult workers. After treatment with propolis at 45 °C, the reduction in weight dropped to zero since the mites died immediately, and, therefore, there was no change in weight. This phenomenon of total mite death was also observed at 40 °C for mites from adult workers, where no change in weight was observed after treatment. The lack of weight change after treatment of mites from adult workers with 4% propolis at 40 °C is in agreement with the absence of heat production, indicating that the mites were dead.

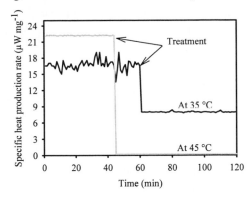

Fig. 4.5 Effect of treatment of *Varroa destructor* mites with 4% propolis on the structure and level of the *p-t* curve, in a typical calorimetric experiment with 30 mites from adult workers at 35 °C and 45°C. A time gap of 30 minutes (omitted in the graph) was required after treatment for the thermal equilibration of the calorimeter.

A typical power-time curve of Varroa mites is usually structured at the normal hive temperature due to mite locomotor activities. But at elevated temperatures the curves are highly smoothed and lie at higher levels, although they last for a short period of time (Fig. 4.5). Treatment of mites with propolis made the *p-t* curves lose their structures and the latter became nearly smooth. Though the specific heat production rate at 45 °C was higher than that at 35 °C, treatment of mites with propolis had a stronger impact on the former, reducing the mean heat production rates by 99.8% and 70%, respectively. In addition to that both curves after treatment were highly smoothed.

Table 4.1 The effect of propolis on Q_{10}
The effect of treatment of *Varroa destructor* mites with 4% Propolis in 55% ethanol on the Q_{10} values. $Q_{10\ Heat}$ and $Q_{10\ Oxygen}$ represent the change in heat production and oxygen consumption rates due to change of temperature by 10 K. 20 to 30 and 50 to 60 mites per calorimetric and respirometric experiments, respectively.

Temperature shift	Before treatment		After treatment	
	$Q_{10\ Heat}$	$Q_{10\ Oxygen}$	$Q_{10\ Heat}$	$Q_{10\ Oxygen}$
25 to 35 °C	2.4	2.3	2.0	1.5
30 to 40 °C	1.1	1.2	1.7	1.4

4.5 Discussion

The higher heat production rate of the phoretic mites at lower temperatures could be an indication that they are adapted to the low temperatures which they are confronted with on the workers' surface during the flying activity of the latter. It is obvious that the phoretic mites are more often exposed to such lower temperatures than mites on brood in capped cells, since the temperature of the beehive is highly regulated, whereas the surface of a flying bee is not. The heat production rate of Varroa mites from adult workers, worker brood, and drone brood grew with increasing temperature, indicating their thermo-conformer physiological nature; the rate remained at a nearly constant level between 30 and 40 °C, with a slight increase between 30 and 35 °C. The constant level of heat production rate between 30 and 40 °C demonstrates that this range is their normal physiological and/or tolerable temperature range. It was demonstrated by Rosenkranz (1985) that Varroa mites prefer temperatures of 34 °C and below. This shows that the higher temperature range of 35 to 40 °C, marked by a similar heat production rate as at the preferred lower temperature range, is tolerated by Varroa mites, though it is not preferred. In addition to this, there is no significant difference in the heat production rate of the mites from the three groups at a particular temperature in this normal physiological temperature range, indicating that they are all equally well adapted. The mites from adult workers, however, had a significantly higher heat production rate at 35 °C.

The very similar heat production rate of mites from drone brood and worker brood in the temperature range of 30 to 40 °C contradicts the idea by Kraus et al. (1998) that Varroa mites reproduce better in drone brood than in worker brood due to the convenient and slightly lower temperature in the former. It is suggested here that the reasons for the higher reproduction rate of *Varroa destructor* in drone brood could be due to other factors, such as the prolonged capped developmental stage of the drone brood. At lower temperatures, however, the mites from drone brood showed a slightly higher heat production rate than those from worker brood. This is easily explained by the fact that the worker brood is usually located at the centre of the comb, with a relatively higher temperature (Kraus et al. 1998), and that the drone brood is located more to the periphery. The very high heat production rate at 45 °C lasted for 90 to 180 min, followed by a sharp decline of the curve, displaying that this temperature is extreme for the mites, and that they were trying to escape, leading to their restlessness and very high metabolic rates. Several researchers used high temperatures to kill mites trapped in worker and drone brood without or with very little damage to the brood (Rosenkranz 1987, Brødsgaard and Hansen 1994, Huang 2001), and also from the surface of adult workers (Hoppe and Ritter 1987). A combined treatment of heat and bee repellent (used to avoid aggregation of bees) produced a strong synergistic varroacidal action (Hoppe and Ritter 1987).

Mites from worker and drone brood displayed similar responses to 4% propolis, but phoretic mites showed a different response, indicating their altered behaviour and/or physiological conditions. The death of phoretic mites after treatment with 4% propolis at 40 °C demonstrates that these mites are highly vulnerable to treatment at this temperature, whereas the mites from worker and drone brood survived with a reduction in their heat production rates by only 68% and 60%, respectively. Treatment of mites at 45 °C resulted in an immediate death displaying the synergistic effect of propolis and temperature.

In addition to displaying the effect of temperature and propolis on the metabolic rate of *Varroa destructor*, the oxygen consumption rates displayed in Fig. 4.3 gives evidence about the length of time one can run the calorimetric experiments with mites, without the need for ventilation. If we consider a mean oxygen consumption rate of 2.0 $\mu l\ mg^{-1}\ h^{-1}$, the total oxygen consumed within 5 h (the maximum experimental time used) will be 10 $\mu l\ mg^{-1}$. Considering 30 mites (ca. 12 mg), the total amount of oxygen consumed during the experimental period is 120 μl. The calorimetric vessel has a volume of 12,000 μl. Since oxygen makes up 21% of atmospheric gas, the total amount of oxygen in the vessel is 2520 μl. During the experimental period, the partial pressure of oxygen in the calorimetric vessel would be reduced from the original 21% to 20%. This shows that the calorimetric experiments, even by using a closed calorimetric vessel, can be run without any problem of oxygen deficiency at least for the first five hours. In addition to this, since the calorimetric vessel is not sealed oxygen can not be a limiting factor.

The treatment of mites with propolis reduced the heat production and oxygen consumption rates proportionally, indicating that the treatment affects both of them, which are directly related to each other only in case of aerobic respiration. The similar values of both curves show that the heat produced during metabolism is due to aerobic respiration. This means that one may use the manometeric method in the metabolic investigation of Varroa mites instead of the expensive calorimeters.

The higher Q_{10} value for the temperature increase from 25 to 35 °C, as compared to that of 30 to 40 °C, before treatment clearly indicates that the metabolism of the mites is well adapted to ambient temperatures around 35 °C which is the temperature found in a beehive. The after treatment Q_{10} values decreased for both the heat production and oxygen consumption rates with the shift of temperature from 25 to 35 °C, as compared to the values before treatment, indicating that the treatment has weakened the mites. In the case of change of temperature from 30 to 40 °C, however, the Q_{10} values after treatment were paradoxically higher than before treatment. The most probable explanation for this could be that the mites from drone brood were agitated due to propolis treatment at 40 °C and that they were trying to escape, or that the higher metabolic rates are indications of metabolic inefficiency, introduced due to propolis poisoning.

It becomes clear from the heat production rate and utilization of reserve food during starvation, that mites from adult workers behave differently than those from the brood stages. The phoretic mites have relatively higher heat production and resource utilization rates at almost all experimental temperatures. This can be due to two reasons: (i) since they are fully grown up they may posses a larger proportion of actively metabolising muscles, contributing to a higher metabolic rate as compared to the mites from the brood stage, which possess a larger proportion of reserve food, and/or (ii) as an adaptation to their way of life i.e. actively attaching themselves to flying bees, in order not to fall down, which needs a larger amount of energy, the phoretic mites might have developed efficient metabolic system.

5.0 Comparative Investigations of the Antivarroa Actions of Propolis from Different Geographic Origins

5.1 Abstract

Varroosis of the western honeybee, *Apis mellifera* L. has become a serious problem for the beekeeping industry in the last 3 decades. Caused by the parasitic mite *Varroa destructor* (Anderson and Trueman), varroosis destroys colonies within 2 to 5 years, unless treated. Due to the emergence of mites resistant to the currently existing acaricides, and the undesirable residue problems caused by the latter, the search for new acaricides is becoming increasingly important. Propolis has already displayed Varroa narcotizing and varroacidal effects in chapters 3 and 4. In this chapter the antivarroa actions of propolis from different geographic origins are investigate *in vitro*, using Petridish bioassay and calorimetric methods.

All propolis samples were extracted in 70% ethanol, and used in 55% ethanol for treatment. In addition to this, 1 sample was extracted, and further used in water. Treatment of mites was carried out by placing them on top of a paper towel in a Petridish, wetted with propolis solution, and keeping them in contact for 30 s.

Treatment with propolis solution resulted in narcosis and the eventual death of mites, regardless of the geographic origin of propolis. The strength of narcosis, illustrated by the time needed for potentially recoverable mites to recover, varied from sample to sample. The lethal effects displayed by a certain concentration of the various propolis samples were not, however, significantly different from each other, regardless of the propolis origin. Sublethal doses of propolis caused drops in the heat production rate and smoothing of the power-time (*p-t*) curves.

The different concentrations of water-extracted propolis showed significantly lower antivarroa actions, demonstrated by the weaker narcotic effects, less mortality of mites, and lower reduction of heat production rates compared to ethanol-extracted propolis samples.

5.2 Introduction

Varroa destructor (Anderson and Trueman) has become one of the most destructive parasites of the western honeybee *Apis mellifera* in the last 3 decades, causing death and withering of several colonies in different parts of the world (Boecking and Spivak 1999). In order to minimize mite infestation rate and subsequently reduce or prevent colony death, beekeepers are using different types of acaricides in the beehive environment.

The widespread and unwise use of synthetic acaricides by beekeepers to combat colony loss has led to the emergence of *Varroa destructor* mites resistant to most of the acaricides in use today in different parts of the world. Mites resistant to fluvalinate and flumethrin have been reported in Europe and the United States (Colombo et al. 1993, Eichen 1995, Lodesani et al. 1995, Milani 1995, Baxter et al. 1998), to coumaphos in Italy and Switzerland (Milani and Della Vedova, 1996), to amitraz in the United States (Elzen et al. 2000), and to bromopropylate and chlordimeform in Europe (Ritter and Roth 1988). In addition to the problem of resistance by mites, acaricidal use is also associated with the contamination of hive products (Kubik et al. 1995, Wallner 1995, Stürz and Wallner 1997, Bogdanov et al. 1998, Wallner 1999), rendering them inconvenient for human use. Honeybee products that suffer from acaricidal contamination are mainly wax and propolis, largely due to the hydrophobic nature of most acaricides.

In order to prevent colony death researchers are searching for acaricides free from the aforementioned problems and are trying to select Varroa-resistant bee races. Though the selection of resistant bee races is a long-lasting solution, it takes longer time. In the meantime, however, colonies have to be treated with acaricides to prevent their loss. In this aspect, the solution which is second to none is the search and use of acaricides free from residue problems. Most of the researches in the area of acaricide screening and search nowadays are concentrating mainly on the use of organic acids and natural products, such as essential oils, because they naturally occur in the beehive and are non-toxic to humans and bees (Kraus et al. 1994). One of these natural products that does not have the problems mentioned for synthetic acaricides, and naturally occurs in the beehive is propolis. Experimental results in the previous chapters displayed that propolis exhibits Varroa weakening and varroacidal actions.

Propolis is a resinous product of the honeybees accumulated in the beehive for different purposes, such as varnish, sealant, putty, bactericide and fungicide. Due to the variation in the types of source plants, propolis samples from various geographical origins have different chemical compositions. This variation may have an impact on the antivarroa actions of the samples. The aim of this investigation was, therefore, to compare the antivarroa actions of propolis of different geographic origins. For this purpose samples were obtained from completely different geographic environments. The detailed mechanisms of action of propolis on Varroa mites shall be investigated using calorimetric methods in addition to the standard Petridish bioassay methods.

5.3 Materials and Methods

5.3.1 Propolis sources

In order to compare the differences in the antivarroa activities of propolis from diverse geographical origins, several samples were obtained from different countries by personal contact with beekeepers and bee researchers in the corresponding countries. The details of propolis sources including the honey bee species that collected them are listed in Table 5.1. Collection of all propolis samples, except the ones bought from local bazaars, according to the information obtained from the donors, was done by scrapping them from the inner wall of the hive and frames of the honey comb. The different propolis samples were acquired as solid raw samples.

Table 5.1 Description of the different propolis samples
Countries of origin, abbreviated names used in the text, colour, and collecting honeybee species of the different propolis samples used. Rus1 was bought from a Russian natural products shop on the "International Green week" exhibition and bazaar (2001) in Berlin. K1 was bought in a bazaar in Almati, Kazakhstan. The exact subspecies of the collecting bees are, therefore, not known. WEP is the water extracted solution of the sample G1.

Geographical Origin	Sample name	Colour	Source Honeybee
Bogota-Colombia	C1	Dark brown	*Tetragonsica angustula*
Holeta-Ethiopia	E1	Dark brown	*A. m. scutellata*
Berlin-Germany	G1	Golden brown	*A. m. carnica*
Berlin-Germany	WEP	Golden brown	*A. m. carnica*
Almati-Kazakhstan	K1	Dark brown	*A. mellifera?*
Russia	Rus1	Greenish brown	*A. mellifera?*
Lublienic-Poland	P1	Brown	*A.m. mellifera*
Opole-Poland	P2	Brown	*A.m. mellifera*
Gradkow-Poland	P3	Golden brown	*A.m. mellifera*
Nysa-Poland	P4	Brown	*A.m. mellifera*
Graham Town–S. Africa	SA1	Dark brown	*A. m. capensis*
Graham Town–S. Africa	SA3	Dark brown	*A. m. capensis*
Graham Town–S. Africa	SA3	Dark brown	*A. m. capensis*
Graham Town–S. Africa	SA5	Dark brown	*A. m. capensis*
Graham Town–S. Africa	SA6	Dark brown	*A. m. capensis*
Graham Town–S. Africa	SA8	Dark brown	*A. m. capensis*
Graham Town–S. Africa	SA11	Dark brown	*A. m. capensis*
Udine-Italy	I1	Brown	*A. m. ligustica*

5.3.2 Preparation of propolis extracts

All samples were extracted in 70% ethanol and used in 55% ethanol for the bioassays. As already described in Chapter 3, 55% ethanol was used as a solvent in the treatment solution to minimize/avoid the effect of strongly concentrated ethanol on the experimental organism. It would have been desirable to prepare different types of extracts from each propolis sample, and to compare their antivarroa activities. However, as the sample sizes of the acquired propolis were too small, it was not possible to do this. Extraction was, therefore, done only with 70% ethanol for all samples but one; a water-extracted propolis (WEP) solution was prepared for the sample from Germany (G1) by extracting it in distilled water, since it was available in the needed quantity. The method of extraction of propolis in water was the same as that used for the extraction in alcohol, with exception of the solvent. This extract was dissolved in distilled water for treatments. Stock solutions (10%) of both the WEP and EEP were prepared in water and 55%

ethanol, respectively, and the treatment solutions, i.e. 2%, 4%, 6%, and 8% were prepared by diluting the stock solutions in the corresponding solvents.

5.3.3 Biological material

The biological material used for this investigation was the honeybee parasitic mite *Varroa destructor* collected from drone brood. As the physiological status and weight of an egg laying and a phoretic female mite could differ, with the weight of the former increasing by up to 30% of that of the latter (Steiner et al. 1995), only mites at the late developmental stages of the drone brood (brown to dark skin pupal stages, which are older than 20 days of brood development) were used in the investigations. Infested drone brood combs around the 20th day of brood development were obtained from the research beehive of the Institute of Zoology, Free University of Berlin. If the brood was not yet in the brown or dark skin developmental stage the comb was incubated further in an incubator at a temperature of 35 °C and RH of 65 ± 5% until the desired stage was achieved. Collection of mites was carried out by opening and inspecting individual brood cells. Any mite obtained from unhealthy (infected) brood was rejected since the physiological status of such mites could be different from those obtained from the healthy (non-infected) ones. During the collection process mites were put in a Petridish containing drone pupae in order to avoid starvation.

5.3.4 Petridish bioassay

The Petridish bioassay was done in order to compare the narcotic effects, and subsequently the lethality of various concentrations of the different propolis samples. Before treatment the mites were put in a Petridish on top of a 3 cm x 3 cm tissue paper (Kimwipes™ Lite 200, Kimberly–Clark™), in order to make handling of the mites easier. Treatment of the mites was done by applying 250 µl propolis solution of the desired propolis type and concentration on top of the tissue paper, not directly on the mites, and keeping them in contact for 30 s. The treatment was ended after the allocated time by removing the tissue paper containing the mites from the Petridish and placing it on a pad of paper towel. The mites were then immediately removed from the treatment tissue paper and placed on a clean paper towel for 1 min to blot the excess propolis solution on their surface. Blotting of the excess fluid from the mites' surface after conclusion of the treatment was especially important in the calorimetric experiments because it otherwise would interfere with the calorimetric signal due to evaporational heat loss of the excess fluid on the mites.

At the end of the treatment the activity of mites was observed by counting the number of narcotized mites at an interval of 30 min for 4 h, starting from the time at which the mites were removed from the paper towel, and placed in a Petridish for further observation. Activity of the mites was examined by gently prodding them with a blunt needle under a binocular microscope. Regardless of the nature of movement, whether only limbs or the whole body, a mite was considered as "active" if it showed even a slight movement, and "narcotized" if there was no movement of any body part. In between the observation times the treated mites were kept in a Petridish containing drone pupae (2 drone pupae per treatment i.e. 6 mites), and placed in an incubator at a temperature of 35 °C and an r.h. of 65 ± 5%. Each experiment was done with 6 mites and repeated 5x.

5.3.5 Calorimetric assay

A 4% w/v propolis solution was selected and used for all calorimetric experiments, since this concentration causes death in some mites and narcotizes and weakens the rest. The calorimeter used was a Biocalorimeter B.C.P.-600 (Messgeräte Vertrieb, München, Germany) with a sensitivity of 44.73 μV mW^{-1} and vessels with volume of 12 cm^3. Each calorimetric experiment was run by using 25 to 30 mites, and was repeated 3 times. The calorimetric signal was recorded using a one-channel recorder with a 1000x built-in amplifier (Kipp and Zonen, The Netherlands).

The heat production rate of mites before treatment was measured for 2 h. Recording was then stopped and the mites were removed from the calorimeter. They were then treated with 250 μl of 4% propolis by keeping them in contact for 30 s, as in the case of treatment with the Petridish bioassay experiment. After ending the treatment the mites were removed from the Petridish and then blotted by putting them on a clean paper towel for 5 min so that no fluid was left on their surface. They were then put back into the calorimetric vessel and the heat production rate was recorded for 2 to 4 h. At the end of the calorimetric experiment the number of dead mites was counted and noted, since it is important to differentiate between the total reduction of heat production rate (due to mite weakening and death), and the reduction of heat production rate of the survivor mites, due to weakening by the treatment.

In order to find out the total heat produced, and subsequently the heat production rate per unit time before and after treatment, the areas below the calorimetric p-t curves were determined using a planimeter (Digikon DK 4261, Kontron Registriertechnik GmbH, München, Germany). The p-t curves were also digitalized and transferred to a PC directly connected to the planimeter. The ASCII data was then imported to statistic and graphic PC programmes for further computation and processing. The results of the calorimetric experiments were presented as specific heat production rates (p, μW mg^{-1}).

5.3.6 Scanning electron microscopy of propolis-treated Varroa mites

In order to evaluate the effect of propolis treatment on the morphology and surface structures female mites were treated with 10% EEP of G1 for 30 s. The surface of the treated mites was then scanned for any deformation or attack by propolis using a scanning electron microscope (FEI, Quanta 200) with a high tension of 10 kV.

5.3.7 Statistical analysis

The statistical tests used for the comparison of the results of different types of treatments were the repeated measures ANOVA of the general linear model (GLM), one-way ANOVA, Tukey's HSD post hoc test, paired sample t-test, and the harmonic mean, according to the nature of the data. Details of matrix construction and analysis method for each statistical test shall be briefly described in the results section up on using the test.

5.4 Results

Treatments with the different types and concentrations of propolis resulted in the narcosis of mites, immediately after treatment, and the activity of mites changed with observation time (Fig. 5.1 a to e). The measurement of mite activity involved repeated observations of the treated mites in time intervals of 30 min, to observe the change in their activity with the length of incubation time after the treatment. The control, a treatment with 55% ethanol, displayed that 90 to 100% of the mites were narcotized immediately after treatment and that narcosis lasted less than 5 min. Since there was no change in the activity of the control treatment mites after 5 min, with 100% of them being active, it was not necessary to display it as a graph with the other treatments, hence, it is missing in Fig. 5.1.

Treatments with 2%, 4%, and 6% WEP had no effect on the activity of mites even immediately after treatment, with all mites remaining active. This was unlike the corresponding weak concentrations of EEP which caused narcosis that changed with observation time. Since there was no activity change of the mites treated with the weak concentrations of WEP mentioned, these concentrations were not displayed in Fig. 5.1 a, b, and c. Though weak compared to the corresponding concentrations of EEP, 8% and 10% WEP showed narcotic and lethal effects that changed with observation time. Treatments with 8% and 10% WEP narcotized 40% and 50% of the mites, respectively, immediately after treatment and these values dropped to

16.7% and 36.7%, respectively, within 30 min. For the treatment with 10% WEP, the narcotic effect on some mites lasted longer and they became active in the time interval of 30 to 60 min, whereas in the case of 8% WEP no activity change was observed after the first 30 min. At the end of the experiment only 16.7% and 27.3% of the mites treated with 8% and 10% WEP, respectively, were dead.

Unlike the treatments with WEP and the control, treatments with different types and concentrations of EEP resulted in narcosis that lasted longer, the length of narcosis being dependent on the concentration of propolis. As it can be seen from Fig. 5.1 a to e, an increasing concentration of propolis resulted in a decreasing number of mites that recovered from narcosis with incubation time (Fig. 1 a versus c, d, and e). Though observation of mite activity was carried out for 4 continuous hours in intervals of 30 min, only the activity results for the first 1 hour and the last observations were presented in the figures. This is because no or very little activity change was observed during the other time intervals. If there was any activity change of mites taking place after the first 60 min observation time, it took place only in the second 60 min interval and remained constant thereafter. With decreasing concentration of propolis the percentage of mites that recovered from narcosis increased

In order to analyse the effect of (i) the type of propolis, (ii) propolis concentration, and (iii) the length of observation time on the activity of mites after treatment, a statistical test called repeated measures ANOVA of the General Linear Model (GLM) was used. This statistical test was run on a matrix containing the type of propolis (17 levels) and concentration of propolis (5 levels) as between subject factors, and the activity status of the mites at 30 min interval observation times (9 repeated observations - levels) as within subject factors. The test was run using a Statistic Programme (SPSS 11.0.1 for Windows), at a significance level of $\alpha = 0.05$. In the case of presence of significant differences within the tested variables, pairwise multiple comparisons were made among the individual levels of factors using the Tukey's HSD post hoc test at a significance level of $\alpha = 0.05$. The results of the analysis demonstrated that there was a significant difference ($p < 0.05$) within the repeated observations (number of active mites) between observations at 0 min and 30 min for all propolis types. Within the next 30 min observation time, a significant number of mites recovering from narcosis was observed at concentrations of 2% and 4% propolis for almost all propolis types (Fig. 5.1 a and b).

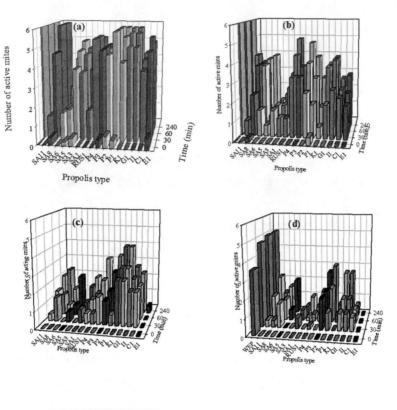

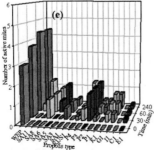

Fig. 5.1 Effect of the ethanol extract of propolis from different geographic origins at concentrations of (a) 2%, (b) 4%, (c) 6%, (d) 8%, and (e) 10% w/v on the activity of *Varroa destructor* mites. The bars indicate mean numbers of actively moving (non-narcotised) mites after treatment of 6 mites per experiment, n = 5. The 2%, 4%, and 6% WEP did not show any narcotic effect, thus, only those of 8% and 10% are displayed (Fig. 5.1 d and e).

At concentrations of 6% and 8% propolis, no significant changes in the activities of mites were observed between the 30th and 60th min of observation time for most propolis types. Significant activity changes for these 2 treatments were observed between the 60th and 90th or 90th and 120th min observation time (Fig. 5.1 c and d). In case of the treatment with 10% propolis, a significant activity change was observed only in the first 30 min; thereafter, practically no mite recovered from narcosis, with the exception of the samples SA5, SA11, P3, K1, G1, and I1, where a statistically significant ($p < 0.05$) activity change was observed between the observation times 30 to 120 min.

In case of the 8% WEP, activity change of the treated mites was observed between the 0 min and 30 min observation time; there was no significant change thereafter. A 10% WEP had a stronger narcotic effect compared to the 8%; hence mites were recovering from narcosis in the observation time interval of 30 to 60 min. There was no activity change observed for both concentrations after 60 min.

In addition to the significant difference at the level of individual between or within subject factors, significant differences also existed at the level of the product of the different subject factors (propolis type x concentration, propolis type x observation time, concentration x observation time, or propolis type x concentration x observation time). In order to discern clearly which group differs from the others, and at which concentration and observation time, a multiple pairwise comparison was done using the Tukey's HSD post hoc test at a level of significance of $\alpha = 0.05$. Following the Tukey's HSD post hoc test, the propolis samples were grouped into homogenous subgroups according to the statistically significant resemblance of the harmonic means of the narcotized Varroa mites (Table 5.2). The harmonic mean is occasionally used when dealing with averaging rates which are difficult to perform with the arithmetic mean, such as in the present case, as described by Croxton et al. (1967).

As can be discerned from Table 5.2, the water-extracted propolis (WEP) segregated from the rest of the group significantly ($p < 0.05$), and lies at the weakest end of the antivarroa spectrum, with a very low mean number of narcotized mites. Even though there was no clear segregation of the ethanol-extracted propolis (EEP) samples into independent groups based on their geographical origins, or bee subspecies that collected them, there was a tendency of the samples from Poland to segregate from the rest, with only the Italian sample (I1), the German sample (G1) and one of the South African samples (SA8) belonging to this group. The 4 South African samples (SA1, SA3, SA5 and SA6) belonged to the same homogenous subgroup together with the samples from Russia (RUS1) and Ethiopia (E1). The weakest homogeneity was observed among the propolis samples from Colombia (C1), 2 from South Africa (SA8 and SA11), 1 from Poland (P4), and 1 from Kazakhstan (K1).

Table 5.2 Grouping of propolis samples into homogenous subgroups based on overall antivarroa activity
Propolis samples were categorised based on the overall homogeneity of the harmonic means of the number of inactivated *Varroa destructor* mites after treatment with various concentrations followed by repeated observation of mite activity. Repeated measures ANOVA, and Tukey's HSD post hoc test ($\alpha = 0.05$) were employed. The numbers in the cells indicate the harmonic means of the inactivated *Varroa* mites by each propolis sample with a 5 (concentrations) x 5 (number of replica) = 25 sample size.

Propolis type	Homogenous subgroups based on $\alpha = 0.05$							
	1	2	3	4	5	6	7	8
WEP	0.62							
G1		3.27						
I1		3.36	3.36					
P2		3.37	3.37					
P1		3.39	3.39					
P3		3.68	3.68	3.68				
SA8		3.73	3.73	3.73	3.73			
P4			3.87	3.87	3.87	3.87		
SA11				4.16	4.16	4.16	4.16	
C1				4.20	4.20	4.20	4.20	
K1					4.31	4.31	4.31	
SA5						4.37	4.37	4.37
SA6						4.39	4.39	4.39
Rus1							4.48	4.48
SA1							4.53	4.53
E1							4.60	4.60
SA3								4.91
Significance	**1.000**	**0.317**	**0.165**	**0.135**	**0.054**	**0.135**	**0.445**	**0.109**

The homogeneity test run for the propolis types, in order to compare and put them into homogenous subgroups, was also run for the different concentrations of propolis to observe if any of the subsequent concentrations have similar antivarroa effects, and form a homogenous subgroup. This test with 17 (propolis types) x 5 (number of replica) = 85 (sample size) for each concentration displayed that no 2 or more concentrations formed a group, and that the antivarroa activity of each concentration was significantly different from that of the others ($p < 0.05$), i.e. no saturation effect was observed.

Further incubation of mites for longer than 4 h displayed no change in the mite's activity, thus, those that were narcotized up to this time were already dead. The number of dead mites at the end of each experiment was counted and presented in Fig. 5.2. Some propolis samples (e.g. G1, C1, P4, and K1) showed similar lethal effects at different concentrations, regardless of differences in geographic origin. Generally, lethality increased with increasing concentration and 10 % EEP killed 85% to 100% of the treated mites.

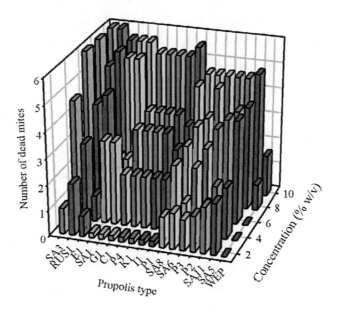

Fig. 5.2 Mortality of *Varroa destructor* mites after treatment with different concentrations of various propolis samples. The bars indicate mean numbers of dead mites at the end of the experiment. Each experiment involved the treatment of 6 mites, n = 5

The treatment of V*arroa destructor* with 4% EEP reduced the heat production rate significantly (paired sample t-test, p < 0.05) by 57.7 to 85.2% for all treatments, except the sample G1 which showed a reduction of 32.8%, still significantly different (p = 0.01) (Fig. 5.3 a). The effect of treatment with G1, displayed by the harmonic mean of percentage reduction in the heat production rate, was significantly lower than the other EEP treatments, and it formed a homogenous subgroup only with the 10% WEP (Table 5.3 b). The 10% WEP reduced the mites' heat production rate significantly (p = 0.02) by 23.2%. The control treatment (55% ethanol) had no significant effect on the heat production rate (p = 0.18) with a reduction by 7.8 ± 2.1%. A one-way ANOVA followed by the Tukey's HSD post hoc test, and categorization of the treatments into homogenous subgroups, based on the significant homogeneity of the harmonic mean reduction of heat production rate due to treatment, is shown in Table 5.3 b. As can be discerned from this table the control segregates from the rest, with a very low and insignificant percentage reduction of heat production rate. WEP and G1 form a subgroup (with a significant resemblance level of p = 0.25) with the lowest harmonic mean among the propolis treatments, indicating the weakness of their antivarroa activity. The rest of propolis samples belong to the same subgroup with a significance level of p = 0.646 and a higher antivarroa activity.

It has to be born in mind that the heat production rate after treatment with propolis is the result of metabolism of the survivor mites. Though it is meaningful to consider the total drop in the heat production rate due to treatment with propolis, since this reflects the extent to which the treatment group as a whole suffered from the treatment, it is also desirable to discern the extent of weakening of the survivor mites. Comparison of the specific heat production rate before treatment with that of only the survivor mites after treatment was conducted by separating the survivors at the end of the calorimetric experiment, and dividing the heat production rate after treatment by the weight of the survivors. Immediately after treatment mites were normally narcotized for a length of time that depended on the type and concentration of propolis, and they produced a negligible amount of heat during this period. During the course of incubation time the mites were recovering from narcosis and started moving, subsequently increasing their heat production rate. The course of heat production rate increased with time until all survivor mites recovered from narcosis, and start metabolizing with the potential they could achieve after treatment. After this point the p-t curve did not ascend anymore and remained at a certain level forming a plateau phase. The heat production rate at the plateau phase was recorded for ca. 2 to 4 h, depending on the stability of level of the p-t curve, after which the experiment was stopped to avoid starvation and death of mites. The survivor mites were separated from the dead ones and weighed. If the experiment has been run longer, the curve would have started declining due to mite starvation and death. Only the heat production rate at the stable plateau phase was considered in the calculation of specific heat production rate of the survivor mites.

As there were no dead mites and hence the number of survivor mites equals the total number of mites after treatment with the control (55% ethanol), the specific heat production rate after treatment was the same whether considered per weight of total number of mites or per weight of the survivor mites. Though there was a drop of the specific heat production rate by 7.8 ± 2.0% due to the control treatment it was statistically insignificant (Paired sample t-test, α = 0.05). Even though the treatments with G1, C1, E1, K1, SA1, SA3, SA5, and WEP caused a reduction of the specific heat production rate when considered per weight of the total number of mites, the specific heat production rate of the survivor mites remained unaffected (Fig. 5.3 b and Fig. 5.4). This indicates that the drop in the specific heat production rate was caused by the death of some mites, but not due to the narcotizing and weakening of the survivor mites.

Table 5.3 Grouping of propolis samples into homogenous subgroups based on the mortality and reduction of heat production rate of *Varroa destructor*
Subgroups of propolis based on treatments with a 4% EEP and a 10% WEP, and observation of (a) the mean percentage of dead mites at the end of the calorimetric experiment, and (b) the percentage reduction of heat production rate. (n = 3, 25 to 30 mites per experiment). The harmonic mean values are displayed. One-way ANOVA and Tukey's HSD test ($\alpha = 0.05$) were employed.

(a)

Propolis type	Subgroups based on $\alpha = 0.05$			
	1	2	3	4
control	0.00			
P2		22.67		
P4		24.11		
WEP		26.2		
G1		33.46		
P1		33.63		
SA5		37.33		
I1		39.13	39.13	
SA8		42.67	42.67	42.67
SA11		43.53	43.53	43.53
SA6		45.83	45.83	45.83
P3		45.55	45.55	45.55
RUS1		49.38	49.38	49.38
K1		54.25	54.25	54.25
C1		56.32	56.32	56.32
E1		58.45	58.45	58.45
SA1				76.29
SA3				78.85
Significance level	**1.000**	**0.074**	**0.099**	**0.083**

(b)

Propolis type	Subgroups based on $\alpha = 0.05$		
	1	2	3
control	7.80		
WEP		23.06	
G1		33.57	
C1			54.67
SA5			56.09
I1			59.33
K1			60.59
SA11			60.65
E1			66.53
P1			69.48
SA8			70.00
P3			71.67
P4			76.59
Rus1			77.45
SA6			77.49
SA3			78.08
SA1			81.59
P2			85.24
Significance level	**1.000**	**0.250**	**0.646**

The treatments with P2 and P4 caused significantly higher reduction of the specific heat production rate of the survivor mites compared to the other treatments (one-way ANOVA and a Tukey's HSD test, p < 0.05), and they formed a homogenous subgroup (p = 0.36). Comparison of the percentage reduction of the survivors' specific heat production rate with that of the percentage mortality data (cf. Fig. 5.4 with Fig. 5.5) displays that the treatments with P2 and P4 caused less mortality. Therefore, their effect is mainly narcotizing and weakening the mites rather than killing them, at least at this concentration.

Comparison of the differences in the percentage reductions of the specific heat production rates achieved by the various treatments showed that there is a statistically significant difference (one-way ANOVA, Tukey's HSD post hoc test, p < 0.05) between 4% EEP and 10% WEP of G1 on one hand, and that of the other EEP treatments on the other hand (Table 5.3 b). The difference between the reductions of specific heat production rate by 4% EEP and 10% WEP of G1 was not significant (p = 0.25), showing the weakness of the water soluble components of propolis; a 10% WEP possessing an activity comparable to a 4% EEP. The percentage reduction of specific heat production rate by the control treatment was significantly lower than the rest of the treatments (one-way ANOVA, Tukey's HSD post hoc test, p = 0.01).

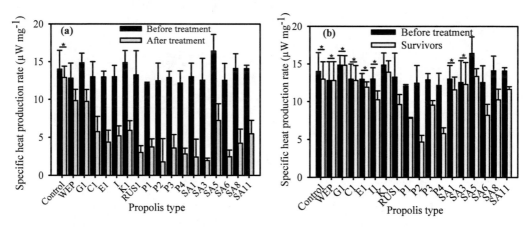

Fig. 5.3 Effect of treatment of *Varroa destructor* mites with a 4% EEP and a 10% WEP on the specific heat production rate of (a) total mites after treatment (b) only the survivor mites after treatment. Mean ± S.D., n = 3, 25 to 30 mites per experiment, * = non-significant difference before and after treatment (paired sample t-test, α = 0.05).

Comparison of the percentage mortality of mites and the percentage reduction of specific heat production rate due to treatment with the different propolis samples displayed that the

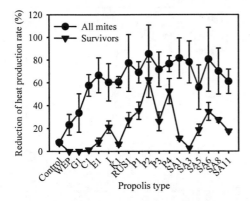 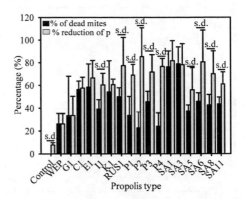

Fig. 5.4 Percentage reduction of mass specific heat production rate calculated based on (a) the weight of all treated mites and (b) the weight of only the survivor mites after treatment with 4% EEP and 10% WEP. n = 3, 25 to 30 mites per experiment, control 55% ethanol.

Fig. 5.5 Comparison of the percentage reduction of heat production rate (p) and percentage of dead *Varroa destructor* mites at the end of the calorimetric experiments after treatment with 4% EEP of different propolis samples and with 10% WEP of G1. n = 3, 25 to 30 mites per experiment. The control involved treatment with 55% ethanol. s.d. = significant difference between percentage reduction of heat production rate and percentage of dead mites (t-test, α= 0.05). ANOVA test results among the different treatments are displayed in Table 5.1 a and b.

percentage of dead mites was either equal to or less than the percentage reduction of the specific heat production rate (Fig. 5.5). In treatments where the percentage of dead mites was not significantly different (t-test, p > 0.05) from the percentage reduction of the specific heat production rate (WEP, G1, C1, E1, K1, SA1, SA3), there was no significant difference between the specific heat production rates before treatment and those of survivor mites (cf. Fig. 5.4 with Fig. 5.5). In such cases, the specific heat production rate of the survivors remained unaffected as before treatment, or it was reduced insignificantly, since the drop in the specific heat production rate was caused by the death of some mites. A one-way ANOVA test on the percentage of dead mites after treatment with different propolis samples followed by Tukey's HSD post hoc test, and grouping the treatments into homogenous subgroups based on the significant resemblance of their harmonic means is displayed in Table 5.3 a. As can be seen, the control does not belong to any other subgroup, due to lack of lethal effect, and the treatments with SA1 and SA3 segregate on the other end of the spectrum with maximum lethality.

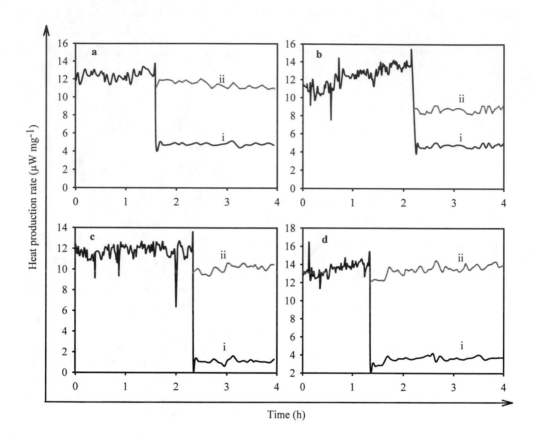

Fig. 5.6 Effect of treatment of *Varroa destructor* mites with 4% EEP of (a) K1, (b) P1, (c) SA1, and (d) SA11 on the structure of the *p-t* curve and on the mass specific heat production rate calculated based on the (i) weight of the total number of mites treated and (ii) weight of only the mites that survived the treatment. 30 mites were used per experiment.

The treatment of mites with 4% propolis not only dropped the specific heat production rate (calculated based on both the weight of total number or only survivor mites), but also changed the structure of the *p-t* curve (Fig. 5.6), smoothing it considerably.

The electron microscopic pictures (Fig. 5.7) show that the treatment of mites with propolis destroyed their cuticle on the different parts of the body, both dorsally and ventrally. The ventral side seems to be highly vulnerable due to the thin layer of cuticle, compared to the thick and more resistant layer on the dorsal side.

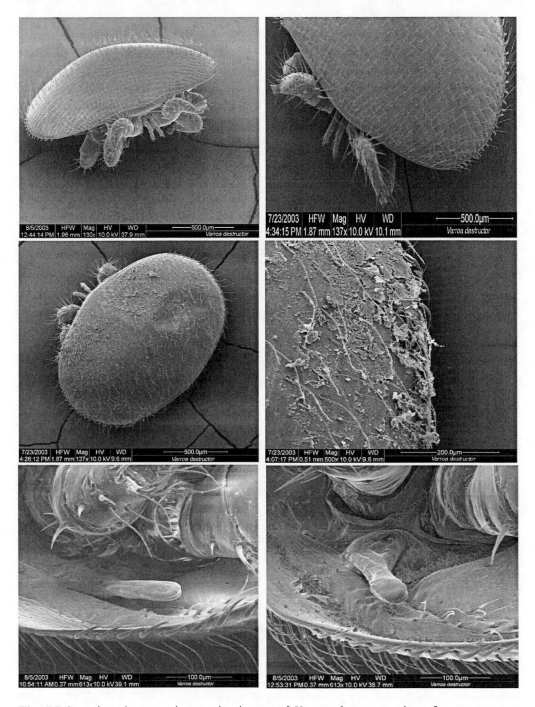

Fig. 5.7 Scanning electron microscopic pictures of *Varroa destructor* mites after treatment with 10% EEP of G1: (a) a control mite, (b) detailed view of the control mite, (c) a treated mite, (d) detailed view of the treated mite, (e) ventral side detailed view of an untreated mite, and (f) ventral side detailed view of a treated mite. (FEI Quanta 200)

5.5 Discussion

The antivarroa action of water-extracted propolis (WEP) was weaker than that of the ethanol-extracted (EEP) ones. The weak biocidal action of WEP compared to EEP was already demonstrated against different types of microbes (Neunaber 1995). The lack of antivarroa action of weak concentrations of WEP (2%, 4%, and 6% w/v) could explain why propolis is not active against mites in the beehive environment. As the WEP makes up a small proportion of the total propolis extract, 3.5% w/w (own result), 2.6 to 6.3% w/w (Spiridonov et al. 1992, Neunaber 1995), 2.0 to 4.0% for Chinese propolis and 6.0 to 14.0% for Brazilian propolis (Miyataka et al. 1997), it is not surprising to find Varroa mites in the beehive, where propolis forms a thin layer on all internal walls and comb cells. The antivarroa actions of higher concentrations of WEP indicate that these components of propolis, in addition to other propolis components such as the volatile ones, could contribute to the Varroa resistance by some bee species. Although the possible insecticidal and varroacidal actions of propolis in the beehive have already been postulated (König and Dustmann 1988, Amrin et al. 1996) until now no experimental evidence existed. If propolis is to play a role in the beehive environment, the 2 groups of substances that are responsible for this action could be the water soluble and the volatile/essential oil components. The Varroa narcotizing and varroacidal actions of WEP ≥ 8% could be proof that in geographical locations where bees collect propolis with higher proportion of water soluble components, propolis could play a role in keeping the Varroa population lower. The majority of components of propolis which are soluble only in non-polar solvents do not have roles in the beehive environment due to insolubility in water.

The strength of narcosis imparted on the mites and subsequently the time needed to recover was dependent on the concentration of propolis. A higher concentration of propolis resulted in a longer time period required for the mites to recover from narcosis, if they were to recover at all. At concentrations above a certain threshold value the antivarroa action of propolis becomes lethal, rather than recoverable narcosis. Strong concentrations, such as 10% E1, Rus1, and SA3 resulted in the immediate death of all treated mites and no mite was observed recovering from narcosis. In addition to variation in the strength of antivarroa effects among concentrations, differences were also observed among samples of different origins at a certain concentration. The difference in the strength of narcosis among different samples could be an indication of quantitative and/or qualitative variations in the narcotic and lethal antivarroa agents they possess. Propolis samples that differ in overall chemical composition may show similar antivarroa effects as far as the substances they possess, which, even if different, have comparable effects.

Based on literature data of propolis from different geographic origins (Marcucci 1995, Kujumgiev et al. 1999, Bankova et al. 2002) it is reasonable to state that the samples differ in their chemical composition. The extent of similarities in the chemical composition of propolis samples depends on the type of plants visited by the bees, which is in turn determined by the geographical location (Ghisalberti 1979, Marcucci 1995, Markham et al. 1996, Miyataka et al. 1997, Burdock 1998, Kujumgiev et al. 1999, Bankova et al. 2000, Bankova et al. 2002). Thus, bees in tropical regions may have a completely different propolis source than bees in the temperate regions. Tropical propolis such as E1 and C1 could, therefore, have completely different chemical composition than temperate propolis. Though the specific chemical make-up of the plant resins collected by the bees differs based on the plant species visited, the general framework of bioactivity could be more or less similar (Kujumgiev et al. 1999), since the purpose for which plants secret resins resemble at least partially. Resin is mainly secreted by plants in order to seal wound, to stop sap loss, to protect wounds from infection by microbes, to protect pollen against infection (it is coated with resin), to stop germination of seeds, and to inhibit sprouting of bud while frost (Ogren 1990). As in case of the use of resin by plants, the use of propolis in the beehive is more or less the same, regardless of the geographic location of the beehives, and this includes defending the hive interior from invading bacteria and fungi and even invading larvae (Ghisalberti 1979, Lisowski 1984, Marcucci 1995).

Even though various samples showed different patterns of activity when considering the length of narcosis, and the number of mites that recover from narcosis in a given period of time, the final effect (lethality) achieved by a certain concentration of most samples was comparable (cf. Fig. 5.1 a to e and Fig. 5.2). This could be an indication of identical functions regardless of origin.

Although the sample G1 did not show significantly lower narcotic or lethal effects at various concentration levels with the Petridish bioassay experiments, its effect on the reduction of specific metabolic rate was inferior to all other EEPs at 4% w/v. This could be due to the fact that the metabolic rate of survivors remains unaffected, apart from the reduction caused by death of certain mites (Fig. 5.3 b).

Comparison of the lethality and reduction of the metabolic rate achieved by a sample shows that the higher the lethality, the higher the reduction in the specific heat production rate, calculated based on weight of the total number of treated mites. In a few cases such as SA1, SA3, K1, C1, G1, E1, and WEP the percentage of dead mites was not significantly different from the percentage reduction of heat production rate (Fig. 5.5). In these cases the reduction of specific heat production rate was mainly due to the death of some mites, but not due to the weakening of the survivors, since reduction of the specific heat production rate of the latter was insignificant and lay near to the zero line (Fig. 5.4). In a number of cases, however, low mortality rates were accompanied by high reductions of the specific heat production rates. This could be due to the fact that the survivor mites were highly weakened and, thus, their metabolic rate was reduced considerably (cf. Fig. 5.4 with Fig. 5.5).

In addition to showing the general antivarroa effect of propolis, which includes death of certain mites and weakening of the rest, the calorimetric method also displays the extent to which the survivors are weakened, and, hence, their metabolic rate is reduced. The treatment with propolis affected the structure of the *p-t* curves, which is the result of the movement and activity of the mites. After treatment with propolis, the *p-t* curve was smoothed (Fig. 5.6) mainly due to the narcotic effect and, thus, the inability of the mites to move actively as they could do before the treatment.

The destruction of the cuticle, creating structural deformities on the mite's protective surface, could facilitate the entry of the biocidal agents into the mite's body, leading to narcosis and consequential death. Since the ventral side is covered with a thin layer of cuticle it can easily be destroyed as compared to the dorsal hard and thick cuticular layer.

6.0 The effect of propolis on the metabolic rate and metamorphosis of the greater wax moth *Galleria mellonella*

6.1 Abstract

Among the moth pests of the honeybee the greater wax moth *Galleria mellonella* (Lepidoptera: Pyralidae) causes the greatest damage, unless controlled at an early stage, because it feeds on wax, pollen, and cocoons of the bee larvae. This leads to the destruction of honeycombs and subsequent deterioration of the weakened colonies. For controlling this pest, natural products are second to none, not least because the use of synthetic substances carries with it the problem of residues which remain in the beehive to affect the bee products. This chapter reports results of calorimetric investigations on the effects of the bee natural insecticidal glue, propolis, on pupal metamorphosis and the metabolic rate of different larval instars.

Experiments were performed by batch calorimetry to record the heat flow rate of individual larvae/pupae before and after the treatment, which consisted of dipping L5, L6, and L7 instars in a graded series of different concentrations of ethanol-dissolved propolis for 30 s before blotting them. The heat production rates were then recorded for 7 h (short period experiment) or during the entire pupal metamorphosis (long period experiment).

The 5th larval instar (L5) showed higher sensitivity to propolis treatment than L6 and L7 whereby total mortality was obtained by 4% propolis for L5 and 8 to 10% for the latter. The treatment of the late L7 stage with non-lethal doses of propolis shortened the duration of pupal metamorphosis significantly. An untreated larva required 6.8 ± 0.8 d (mean $\pm$ SE, n=5) between larval-pupal and pupal-adult ecdysis, whereas this time was shortened to 5.4 ± 0.9, and 4.8 ± 0.5 d after treatment with 1% and 2% propolis, respectively. Though all treated larvae went through larval-pupal ecdysis, 40% and 100% of those treated with 2% and 4% propolis, respectively, displayed abortion of pupal metamorphosis and died. These results indicate that propolis is toxic at higher concentrations and acts as an insect growth regulator at lower ones.

6.2 Introduction

Among the wax moth pests of the honeybee the greater wax moth *Galleria mellonella* L. (Lepidoptera: Pyralidae) causes the greatest damage, leading to material and financial losses. The larval stage of *G. mellonella* (with its 7 instars), the only feeding stage with the longest life span of all developmental stages, builds its silk-lined feeding tunnel in the honeycomb and feeds on wax, pollen, faeces and cocoon of the bee larvae. This voracious nature of the larva leads to the destruction of the honeycomb and the subsequent death of weak colonies. Adults do not feed because they have atrophied mouth parts (Charrière and Imdorf 1997).

The greater wax moth can be controlled by biological, physical, and chemical methods, but most of these methods are either inefficient or expensive for the small-scale beekeeper. In addition, most chemical methods are associated with residue problems in honeybee products. The most commonly used biological control methods include: (i) the control with the wasp *Trichogramma spp.*, that infests the eggs of the wax moth resulting in the emergence of a wasp rather than a moth from the egg of the latter (Bollhalder 1999); (ii) toxin-containing spores of the bacterium *Bacillus thuringiensis,* the toxins being released after spores are ingested by the moth larvae, subsequently damaging the intestinal wall and killing the latter; (iii) the use of male/female moth pheromone (Fraser 1997) to lure and trap the female/male moth; and (iv) using sterile male release technique, this method has been effective under laboratory conditions (Caron 1992). The use of pheromone traps is not as such effective since males find their partners not only by chemical means but also by the use of ultrasound (Caron 1992). Physical methods of wax moth control include the storage of combs at low temperature (<15 °C), frost treatment, and heat treatment (Charriére and Imdorf 1997). Each of these methods, however, has its own drawbacks and is too expensive to use by most small-scale beekeepers. Chemical control methods are also used to prevent or stop the destruction of honeycombs by wax moths. The most commonly used chemical fumigants include sulphur dioxide, acetic acid, formic acid, and paradichlorobenzole (PDCB). Even though these chemicals can be used in case of emergency they have their own limitations in that a number of them cause irritation and poisoning of bees and human beings. Some chemicals used for wax moth control, such as PDCB, contaminate honeybee products, mainly honey and wax (Wallner 1991). Storage of combs in modified atmospheres such as high partial pressure of CO_2 (Yakobson et al. 1997) can also be applied, but mainly by large scale beekeepers.

An alternative, and most likely the best solution, to solve the residue problems associated with chemical treatments, and the financial costs incurred by physical control methods, would be to use natural products that are at hand to the beekeeper and free of the aforementioned problems. One such honeybee product is propolis.

Though propolis is found inside the beehive, it does not play a significant role against the parasites, pests, and pathogens of the honeybee *in situ*. However, the *in vitro* experiments, presented in the previous chapters (3 to 5), demonstrated that propolis is varroacidal. The findings of a number of researchers confirmed that propolis is bactericidal and fungicidal against pathogens of the honeybee that cause disease such as foulbrood, chalkbrood, and others (Lindenfelser 1967, König and Dustmann 1988).

The potential residue free use of propolis, compared to the commercially available expensive, hazardous and residue-associated insecticidal agents employed in the combat against *Galleria mellonella*, provide an incentive to investigate its insecticidal action against this wax moth. The aim of this chapter is to demonstrate the insecticidal and/or insectistatic (abort insect larval/pupal development) actions of propolis.

Calorimetric techniques were used for the various investigations. Calorimetry is a useful tool in the continuous monitoring of different developmental processes throughout the whole life cycle of individual insects without interference in their normal physiological activities. Several researchers have employed calorimetry in the investigation of insect growth and development; among others for the well studied *Galleria mellonella* by Löhr et al. (1978), Schmolz and Schulz (1995), Harak et al. (1996), Lamprecht (1997), Lamprecht (1999), Schmolz and Lamprecht (2000). Moreover, the insect growth regulator and toxic effects of plant secondary metabolites on insects have been investigated calorimetrically (Kuusik et al. 1995). Standard bioassay methods, such as the Petridish bioassay, demonstrate the results of extreme cases of biological activity, such as lethality of a certain concentration, or its impotence, demonstrated by the survival of the organisms after treatment. Biological activities of sublethal concentrations and their effects on development of organisms could, however, be monitored online by the use of the calorimetric method. This latter method was found to be highly sensitive in the investigations of effects of plant secondary metabolites on insect metamorphosis (Kuusik 1995) and on the Varroa weakening action of propolis demonstrated in the previous chapters (3 to 5). The calorimetric method is highly sensitive since it monitors the heat generation rate of an organism, which is directly determined by the metabolic rate. In addition to its high sensitivity, the calorimetric method enables the researcher to judge the mode of action of an insecticidal/insectistatic agent.

6.3 Materials and Methods

6.3.1 Animal and culture conditions

The greater wax moth *G. mellonella* was cultured in a plastic container (25x25x10 cm) at ambient temperature of 30 °C, relative humidity of ca. 70% and 24 h darkness. The culture medium (larval food) consisted of 22% maize flour, 11% wheat flour, 11% bruised wheat, 11% milk powder, 5.5% yeast, 17.5% beeswax, 11% honey, 11% glycerine. All the larval stages and eggs were kept together separated from the pupal and adult stages.

As the early larval stages are too small to handle and too delicate to be used for the purpose of the present investigations, only the 5th (L5), 6th (L6), and 7th (L7) larval instars were chosen. Identification of each larval instar was made by the width of the head capsule and its weight as parameters (for details, see Sehnal 1966). Only larval instars with weight and head capsule width values nearest to the mean of the corresponding range were considered.

6.3.2 Calorimetric experiments

The calorimetric experiments were performed using 3 isoperibolic heat conduction batch calorimeters with different vessel volumes. All investigations with L5 were conducted with a Biocalorimeter B.C.P.-600 (Thermanalyse, München, Germany) with a vessel volume of 12 cm^3, and a sensitivity of 44.73 µV mW^{-1}. For the corresponding experiments with L6 and L7 larvae, 2 Calvet calorimeters (SETARAM, Lyon, France) with vessel volumes of 15 cm^3 and 100 cm^3, respectively, were used. Each of these calorimeters has 2 measuring and 2 reference vessels. The sensitivities of the instruments amounted to 62.63 and 44.21 µV mW^{-1} for the 2 vessels with volume of 15 cm^3 and 51.15 and 53.67 µV mW^{-1} for the 2 vessels with volume of 100 cm^3.

In order to avoid starvation and behavioural change the larvae were provided with sufficient food for the entire experimental period. Air exchange between the vessel content and the surrounding took place through the openings in the lid of the Pyrex glass vessels. 2 types of calorimetric investigations were performed: short and long time experiments.

6.3.2.1 Short period experiments

The aim of these experiments was to investigate the effect of different sublethal concentrations of propolis on the heat production rate of the 3 larval instars mentioned (L5, L6, and L7), and to compare the change in the sensitivity to propolis, if any, with changing larval instar. Both the measuring and reference vessels were supplied with equal amounts of food to avoid asymmetry of non-experimental factors in the 2 vessels. The presence of the larval food in the vessels does not affect stability and level of the baseline. After establishment of the baseline, a pre-weighed larva was placed into the measuring vessel, and the heat production rate was recorded for ca. 4 h. Then the larva was removed from the vessel and treated with propolis, as described below. The treated larva was put back into the calorimeter and the heat production rate was again recorded for 7 h. Each experiment was done 6 times and results are presented as mean ± SE.

6.3.2.2 Long period experiments

These experiments were conducted only with L7. The aim was the evaluation of the effect of sublethal concentrations of propolis on metamorphosis and development of the pupal stage. This could answer the query whether sublethal concentrations of propolis, without remarkable effects on the larva, could cause abortion of pupal development, or either shorten or prolong the pupal development time. The heat production rate of the untreated, pre-weighed larva was recorded for a day in order to observe its activity before treatment. The larva was removed from the calorimeter, weighed again, treated with the desired propolis concentration and put back into the calorimeter. The heat production rate was recorded further until adult emergence, with weight measurements every 24 h. The mean weight between 2 consecutive weighings was used in the calculation of the specific heat production rate in this period (24 h). In cases where there was no adult emergence recording was continued for a total of 25 to 30 days and finally the calorimetric vessel was opened. The pupa was then removed and inspected for life under a binocular microscope by pricking it with a blunt needle. The maintenance of a constant weight during pupal development was also used as a preliminary clue for the death of the organism. Each treatment, including the controls, was done 5 times and the values are presented as mean ± SE.

6.3.3 Propolis preparation and larval treatment

Propolis samples obtained from the research beehives of the Institute of Zoology, Free University of Berlin, were extracted in 70% ethanol. From the extracted and dried propolis sample, a 10% w/v (g ml^{-1}) propolis stock solution was prepared in 55% v/v ethanol. The desired concentrations for treatment (0.25, 0.5, 1.0, 2.0, 4.0, 8.0, and 10.0 % w/v) were obtained by diluting the stock solution with 55% ethanol.

Treatments were made by dipping the larvae in 5 ml propolis solution in a 30 ml vial for 30 s. After the allocated treatment time the larvae were removed with a pair of tweezers taking care not to damage them, and placed on a pad of absorbent paper towels for 1 min, to blot fluid, which would otherwise disturb the calorimetric signal and prolong the experimental time undesirably. Double control experiments were carried out by dipping the larvae in 55% ethanol and in distilled water. After being properly blotted the larvae were put back into the calorimeter. Recording of the heat production rate started after a thermal equilibration time of 30 to 45 min, which is always needed after replacing the calorimetric vessel.

6.4 Results

Unless it is clearly stated, all values in this work are given as mean ± SE.

6.4.1 Metabolism and growth of untreated larvae

The wet weight of the larval stages increased drastically, from a mean value of 23.0 ± 2.5 mg at L5, achieving its maximum mean value of 236.8 ± 48.1 mg at the 7th larval instar (Fig. 6.1). These values are means of the larval instars used in the present investigation. Otherwise, the weight change during the entire larval developmental stage ranges from < 1 mg for L1, to nearly 400 mg at the 7th larval instar of some individuals. This drastic increase in weight was then followed by a nearly uniform drop during pupal metamorphosis, as it will be seen in the next sections.

The mean total heat production rates of untreated larvae increased with larval age from L5 (1.7 ± 0.2 mW) to L7 (6.5 ± 0.4 mW) and dropped drastically at the pupal stage (1.9 ± 0.3 mW). The specific heat production rates, however, followed a reverse pattern, except at the pupal stage, dropping considerably from L5 (78.9 ± 8.9 mW g^{-1}) to the pupal stage (8.2 ± 3.1 mW g^{-1}) (Fig. 6.2).

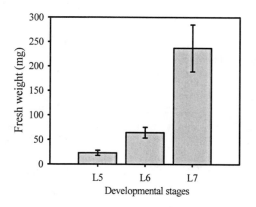
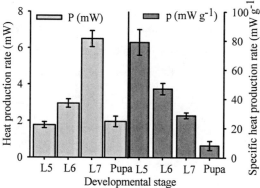

Fig. 6.1 Mean fresh weight (mg) ± SD of the 5th, 6th and 7th instar larvae of the greater wax moth *Galleria mellonella*. n = 54 for L5 and n = 36 for L6 and L7 each.

Fig. 6.2 Heat production rate (*P*, mW) per individual larva and specific heat production rate (*p*, mW g^{-1}) of L5, L6, L7 and pupa of the greater wax moth *Galleria mellonella*. L5, L6, and L7 represent the 5th, 6th, and 7th instar larvae, respectively. Mean ± SE, n = 54 for L5 and pupa and n = 36 for L6 and L7 each.

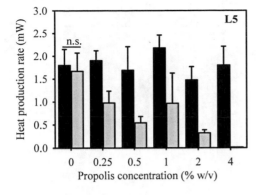

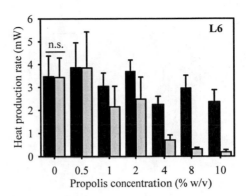

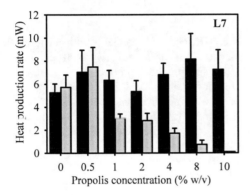

Fig. 6.3 Effect of treatment with different concentrations of propolis, in 55% ethanol, on the heat production rate (per animal) of the different instar larvae: L5, L6, and L7 of the greater wax moth *Galleria mellonella*. Mean ± SE, n = 9 (for L5) and n = 6 for L6 and L7. n.s. = no significant difference (paired sample t-test, p> 0.05).
■ Before treatment
□ After treatment

6.4.2 Short period experiments

The 5th larval instar was highly sensitive to propolis treatment compared to the 6th and 7th instars. Whereas the 2 latter instars did not display sensitivity even to 0.5% propolis, the heat production rate of the 5th larval instar was reduced by 48% due to treatment with 0.25% propolis (Fig. 6.3). Treatment with 4% propolis resulted in 100% mortality of L5 and reduced the heat production rate of L6 and L7 to 25% - 30% of the initial values (Fig. 6.3). Although the heat production rate of L5 dropped by 7.4%, this change was not statistically significant (paired sample t-test, p = 0.09); and the other 2 larval instars did not show any observable sensitivity to the control treatment (Fig. 6.3). The 6th and 7th larval instars were sensitive to treatments with propolis concentrations of ≥ 1%, with 10% propolis resulting in 100% mortality of L7 and a decrease of the power-time *(p-t)* curve to the baseline. The same concentration reduced the heat production rate of L6 by 95% (Fig. 6.3 b and c). In addition to the change in the total heat production rate, the change in the specific heat production rate showed a similar pattern (Fig. 6.4). The dose-response curves of residual specific heat production rate versus propolis concentration were the same for L6 and L7, but L5 displayed a different pattern (Fig. 6.5).

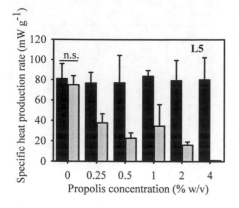

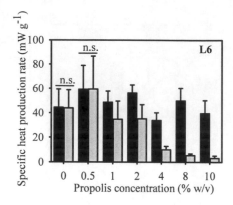

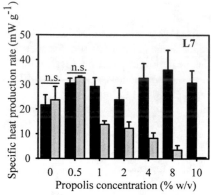

Fig.6.4 Effect of treatment with different concentrations of propolis, in 55% ethanol, on the specific heat production rate of the different instar larvae: L5, L6, and L7 of the greater wax moth *Galleria mellonella*. Mean ± SE, n = 9 (for L5) and n = 6 for L6 and L7. n.s. = no significant difference (paired sample t-test, p > 0.05).
■ Before treatment
☐ After treatment

6.4.3. Long period experiments

The typical (control) long period *p-t*, curve of *G. mellonella* development in the last larval and the pupal stage showed a drastic drop in the heat production rate from the late L7 to the pupal stage through the prepupa (late L7 enclosed in silk cocoon) (Fig. 6.6 a).

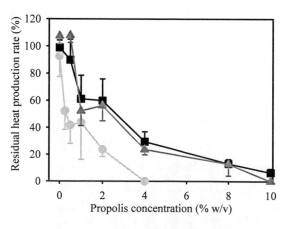

Fig. 6.5 Percentage residual specific heat production rates of the 5th (L5), 6th (L6) and 7th (L7) larval instars of the greater wax moth *Galleria mellonella* after treatment with various concentrations of propolis in 55% ethanol. Mean ± SE, n = 9 (L5) and n = 6 (L6 and L7).

● L5
■ L6
▲ L7

The transition from prepupa to the pupal phase around day 2 was accompanied by a sharp peak followed by a trough at ecdysis. The pupal heat production rate then dropped from a mean value of 2.2 ± 0.9 mW to 0.9 ± 0.3 mW in 1 day and remained at this level for the next 3 days. The heat production rate started to ascend between the 3rd and 4th days after pupation and achieved a maximum value of 3.6 ± 0.8 mW on the 6th day after pupation. The pupa-adult moulting took place 6.6 ± 0.7 days after the larva-pupa moulting. This last moulting was accomplished after a strong muscular contraction activity displayed by a sharp peak of 8.2 ± 0.45 mW followed by a trough of 0.2 ± 0.1 mW (Fig. 6.6 a).

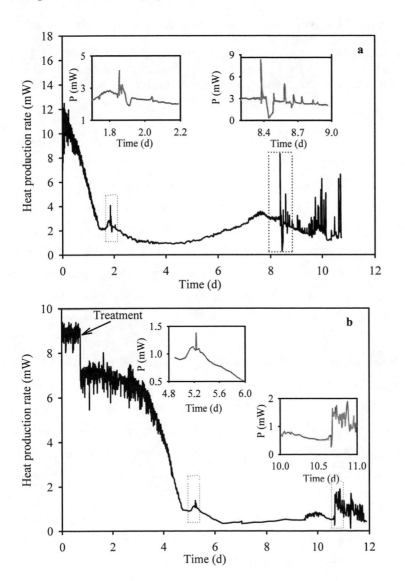

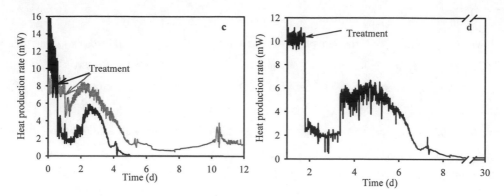

Fig. 6.6 Typical power time (*P-t*) curves of the development of *Galleria mellonella* from the late 7th larval instar to adult emergence. (a) 307 mg larva without treatment, (b) 173 mg larva treated with 1% propolis, (c) 209 mg (i) and 308 mg larvae (ii) treated with 2% propolis, and (d) 189 mg larva treated with 4% propolis. Note the differing vertical scales. Treatment period was 30 s. The insets are enlarged portions of the larval-pupal and pupal-adult ecdysis, marked by rectangles on the curve underneath the corresponding inset.

The 5 pupae treated with 1% propolis during the late L7 stage successfully completed their development to adult emergence. A representative example of pupal development after treatment with 1% propolis is displayed in Fig. 6.6 b. Though some of the larva-pupa and pupa-adult ecdyses were accompanied by peaks and troughs on the *p-t* curves, they were much weaker than the control peaks and near to the "resting" heat production rates. Some of the ecdyses displayed only exothermic peaks but no troughs (Fig. 6.6 b). The pupal-adult moulting of these treated organisms showed a unique feature in that there was no sharp exothermic peak followed by a trough, unlike the controls; rather the moulting was displayed by raising the level of the curve to a higher value. The pupal metamorphosis lasted 5.38 ± 0.9 d, which is shorter than that of the control.

Among the 5 larvae treated with 2% propolis, only 3 completed pupal development whereas 2 of them, even though they accomplished the larva-pupa ecdysis, were unable to complete their pupal development. The peaks and troughs during ecdyses were not as strong as in the case of the controls. In addition to that, the emerged adults, like the ones of 1% propolis treatment, did not show the typical *p-t* curves displayed by the controls; it was rather weak locomotory (flying) activity (cf. Fig. 6.6 a, b, and c). The pupal metamorphosis after treatment with 2% propolis lasted 4.8 ± 0.5 d.

Though the larvae treated with 4% propolis performed the larva-pupa ecdyses, the pupal development was aborted in all the 5 larvae investigated. These results indicate that, though the larvae survived the treatment and had a residual heat production rate of ca. 25% (Fig. 6.5), they were too weak and too unhealthy to go through pupation.

Apart from the difference in the heights of peaks (exothermic) and depths of troughs (endothermic) associated with moulting; the treatment with different concentrations of propolis introduced a significant difference in the length of the pupal metamorphosis (Table 6.1).

Table 6.1 Effect of propolis on length of pupal metamorphotic phase
After the late L7 stages of the greater wax moth *Galleria mellonella* were treated with sublethal concentrations of EEP, they were allowed to go through metamorphosis. The length of time (days) needed to complete this phase was recorded. A 1-way ANOVA and the Tukey's HSD post hoc test ($\alpha = 0.05$) were employed. Tukey`s test results with identical letters show no significant difference, but different letters do. Mean $\pm$ SE, n = 5.

Treatment	Length of pupal metamorphosis (d)	Tukey`s test
Control (no treatment)	6.6±0.7	a
Control (55% ethanol)	6.8±0.8	a
1% propolis	5.4±0.9	b
2% propolis	4.8±0.5	b

The specific heat production rate during pupal development showed a typical U-shaped curve for the controls. The curves for the treated pupae were, however, flatter with a smaller heat production rate in the late pupal and adult stages (Fig. 6.7).

The change of weight during the pupal development displayed similar patterns regardless of the treatment (Fig. 6.8). All pupae investigated showed a uniform loss of weight with developmental time until adult emergence.

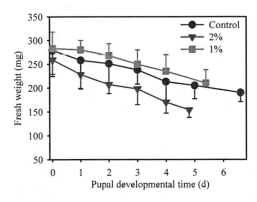

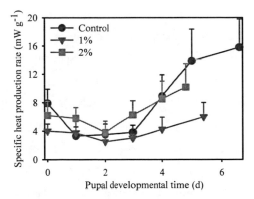

Fig. 6.7 Specific heat production (mW g^{-1}) of the greater wax moth *Galleria mellonella* during pupal development after treatment of the 7th instar larva with different concentrations of propolis. The pupal developmental time was counted starting from the larval-pupal moulting day as zero. Mean ± SE, n = 5.

Fig. 6.8 Loss of weight of the greater wax moth *Galleria mellonella* during pupal development after treatment with different concentrations of propolis at the 7th larval instar. Only pupae that successfully completed pupal development were considered. The day of larval-pupal moulting was regarded as day zero. Mean ± SE, n = 5.

6.5 Discussion

The wet weight of larvae increased exponentially from the 5th to the 7th larval instar and dropped uniformly during pupal metamorphosis. This is because the larval stage is a "feeding machine", continuously consuming available food in order to accumulate enough reserve food for the entire phase of pupal metamorphosis and for the flying and reproductive activity of adults. As the pupal phase does not feed and hence completely depends on the reserve food accumulated during the feeding larval stage, its weight decreases continuously and at constant rate during metamorphosis. The uniform drop in the weight of the pupa during metamorphosis indicates that the rate of utilization of reserve food during this phase of *Galleria* development is constant. The tissue composition (proportion of fat, proteins and carbohydrates) remains almost constant at the various larval and pupal stages, changing only in the adults with increasing fat proportion (Schmolz et al. 1999).

The heat production rate increased from 1.77 ± 0.17 for L5 to 6.51 ± 0.44 for L7 mainly due to the increase in the wet mass. However, the specific heat production rate decreased from 78.8 ± 8.8 for L5 to 28.6 ± 2.0 for L7 and 8.2 ± 3.1 for the pupal stage. This decrease in the specific heat production rate is mainly because the bulk of the increased weight of L6, L7 and the pupal stages is reserve food and not metabolizing tissue. In the 5th larval stage, the main component of the larval weight is metabolizing tissue, leading to a very high specific heat production rate (Schmolz and Schulz 1995, Schmolz et al. 1999).

The higher sensitivity of the 5th larval instar to propolis treatment is due to at least 2 factors: the thin and relatively permeable cuticular layer, leading to a greater penetration of propolis; and the very high specific heat production rate, incorporating the toxin into the metabolic machinery at a faster rate.

Since the larval cuticle, or exoskeleton, stretches only to a limited extent it must be shed periodically to accommodate the rapidly growing body size of the larva, which could double daily during the first 10 days under ideal conditions (Morse 1978). Though the basic outer layers of the new cuticle are formed before shedding the old one, additional layers of endocuticle are added and sclerotization of outer layers increases with developmental days, throughout the duration of the instar (Semple et al. 1992). This indicates that the strength and thickness of the cuticle increases with age of the larval instar. The life spans of L5, L6, and L7 under ideal conditions are 2.2, 3.0, and 7.5 days, respectively (Sehnal 1966). Thus, L5 has the thinnest cuticle with high permeability and L7 has the thickest cuticular layer that impedes penetration of the lipophilic components of propolis. It is therefore highly plausible to state that, the thickness of the larval exoskeleton plays a role in the insecticidal action of propolis.

The higher specific heat production rate of the L5 plays a role in the faster penetration of propolis across the cuticle, and its accumulation at higher concentrations in the tissue. Higher metabolic rates are associated with increased transport of hydrocarbons and lipids through lipid pore canals across the cuticular layer (Renobales et al. 1991). The transport of hydrocarbons and lipids across the lipid pore canals to the cell surface may allow inward transport (penetration) of nonpolar pesticides, as displayed by direct correlation between active biosynthesis of hydrocarbons and transport to the surface and penetration by pesticides, especially the nonpolar ones (Theisen et al. 1991). As the majority of bioactive components of propolis are nonpolar, the analogy of penetration of nonpolar pesticides and propolis across the cuticular layer is reasonable. In addition to penetration via transport mechanisms, propolis may attack the cuticle aggressively and penetrate by destroying underlying structures.

When moulting, the cuticle begins to separate from the epidermis, the larva reduces feeding activity and becomes quiescent. Each active stage in the larval life is thus followed by a sluggish premoulting period (Snodgrass 1935). This quiescent stage in the transition from larva to pupa is accompanied by the declining heat production rate and a "U-shaped" *p-t* curve shortly before larval-pupal ecdysis. In this quiescent phase, a part of the old cuticle is degraded, resorbed and recycled by the epidermal cells for the formation of the new cuticular layer. Final break up of the old exoskeleton is achieved by peristaltic contraction of abdominal muscles, raising blood pressure in the thorax and splitting the former at the weakest point, usually along the mid dorsal line (Semple et al. 1992). This contraction of abdominal muscles is accompanied by a sharp peak on the *p-t* curve, and the break of the old cuticle and subsequent release of exuvial fluid is shown by the trough of evaporational heat loss. The height and area of the sharp peak and the depth of the trough indicate the amount of energy spent on contraction of the muscles and evaporation of exuvial fluid, respectively.

The treatment with propolis disturbs the above described typical moulting activity features of *Galleria mellonella*. After treatment with 1% propolis, all larvae were able to go through the metamorphotic phase and emerge as adult, but the exothermic and endothermic peaks were smaller than those of the controls. The adult emerged after the unusual moulting behaviour displayed a very weak flying activity demonstrated by the form of the *p-t* curve (cf. Fig. 6.6 a with b and c). The weak flying activity could be the result of malformed/underdeveloped flying muscles and or wing structures. This could be plausible as the metamorphotic phase/morphogenetic stage is significantly shortened due to treatment, ending up in the improper formation/deformation of body structures.

With the increase in the concentration of propolis, the length of the pupal phase was shortened significantly from 6.8 ± 0.8 d (ethanol control) to 4.8 ± 0.5 d (2% propolis) (Table 6.1). This suggests that propolis accelerates the development of the larval/pupal stage of *Galleria mellonella*. The unusually higher rate of metamorphosis may lead to malformed and immature individuals.

The biological activity of propolis displayed on *Galleria mellonella* is comparable to that of insect growth regulators (IGR) and toxicants, calorimetrically investigated by several researchers. Among others, Kuusik and colleagues (Kuusik et al. 1993, Kuusik et al. 1995) and Harak et al. (1999) elucidated that IGR and toxic compounds/mixtures interfere with the form of the *p-t* curve of insect development, even leading to abortion of metamorphosis. It was stated (Strong and Dickman 1973) that IGR could act to inhibit, retard or even accelerate insect developmental processes. The biological activity of propolis on *Galleria mellonella* obtained in the present investigation fits with those that accelerate insect development. It was also demonstrated by several researchers (Snodgrass 1935, Williams 1967, Strong and Dickman 1973, Theisen et al. 1991, among others) that the application of IGR at the larval stage resulted in the disruption of pupal development and early adult emergence. In addition, normal ecdysis was not achieved. Additionally, treated larvae may give rise to morphologically deformed adults that are unable to fly properly (Metwally and Sehnal 1973).

The use of moderate concentrations of propolis, such as 4%, in the control of *Galleria mellonella* is reasonable since it is toxic and kills the early larval stages immediately, facilitates larval-pupal ecdysis, and aborts pupal development. The practical significance of such concentrations of propolis is that they help to avoid the use of higher propolis concentrations that could probably affect the quality of honeybee wax and also avoid unnecessary wastage of propolis. Propolis can naturally occur in beeswax to a certain degree, but higher concentrations may be undesirable in some uses of wax, such as in the cosmetic industry where propolis may cause allergy in very few individuals.

7.0 Microbiological and calorimetric investigations on the antimicrobial actions of different propolis extracts: an *in vitro* approach

7.1 Abstract

Propolis is a honeybee product which has been used by humans since ancient times as a multifaceted antimicrobial drug against several types of ailments. It had been, however, forgotten in the meantime due to the discovery and effective use of antibiotics. Nowadays the medicinal application of propolis is reviving mainly due to the emergence of pathogenic bacterial and fungal strains resistant to drugs, the undesirable side effects of some synthetic drugs, and also to the "back to nature" trend. Even though several publications exist in the realm of research on the antimicrobial actions of propolis, most of them are limited in scope by the use of only 1 type of extract, of samples from only 1 geographic location, and the inability to discern the mechanism of antimicrobial action.

In this chapter, the antibacterial action of 3 different types of extracts of propolis: (i) water-extracted propolis (WEP), (ii) propolis volatiles (PV), (iii) and ethanol-extracted propolis (EEP) were investigated by flow microcalorimetry coupled with polarography, and by Petridish bioassay methods. Antifungal investigations were performed only with the Petridish bioassay method because of difficulties with growing fungi in flow calorimeters. Antimicrobial activities were compared by the minimal inhibitory concentrations (MIC), minimal bactericidal concentrations (MBC), some calorimetric power-time curve characteristics, and the mean diameter of inhibition zone in Petridish bioassays.

The water-extracted propolis solution had the weakest antibacterial and antifungal action, compared to the other 2 extracts, which showed effects nearly similar to each other. The filamentous fungi were generally less sensitive to propolis than bacteria and yeasts, regardless of the types or concentrations of propolis. Treatment with some propolis samples at concentrations around or lower than MIC values, stimulated effluent growth of *Pseudomonas syringae*, a phenomenon called hormesis.

Flow microcalorimetric and polarographic investigations of the antimicrobial actions of propolis displayed its mechanism of action; both bacteriostatic and bactericidal actions were displayed depending on the concentration, type of propolis, and type of bacteria tested. The Gram negative *E. coli* was insensitive to most treatments, and higher concentrations of propolis, than for other bacteria, were required to achieve bactericidal effects.

Treatments of bacteria with weak propolis concentrations dropped the calorimetric power-time (*p-t*) curves to lower levels, where the curves remained for the rest of the experimental period, or dropped to the baseline with the course of time, or revived after some time and attained peaks. The treatment with strong concentrations, however, dropped the curves to the baseline immediately.

7.2 Introduction

Propolis has been used by man since early times, for various purposes, especially as a medicine because of its antimicrobial properties (Crane 1990, Cheng and Wong 1996). Ancient Greek texts refer to the substance as a "cure for bruises and suppurating sore", and in Rome propolis was used by physicians in making poultices. The therapeutic properties of propolis (tzori, in Hebrew) are mentioned throughout the Old Testament. Records from 12th century Europe describe medical preparations using propolis for the treatment of mouth and throat infections, and dental cares (Krell, 1996). Several antimicrobial activities have been ascribed to propolis including antibacterial (Metzner et al. 1977, Grange and Darvey 1990, Ikeno et al. 1991, Kujumgiev et al. 1993, Aga et al. 1994, Digrak et al. 1995, Garedew and Lamprecht 1997, Menezes et al. 1997, Kujumgiev et al. 1999, Santos et al. 2002), antifungal (Pepeljnjak et al 1982, Dobrowolski et al. 1991, Digrak et al. 1995, Otta et al. 2001, Sawaya et al. 2002), antiprotozoan (Scheller et al. 1977, Starzyk et al. 1977, Dobrowolski et al. 1991, De Castro and Higashi 1995), and antiviral (König and Dustmann 1988, Amoros et al. 1992), among others.

Most of the hitherto antifungal investigations of propolis concentrated mainly on yeast, such as different species of *Candida,* or dimorphic human and animal pathogenic fungi (Lindelfelser 1967, Pepeljnjak et al 1982, Dobrowolski et al. 1991, Digrak et al. 1995, Otta et al. 2001, Sawaya et al. 2002) with satisfactory fungistatic and fungicidal results both *in vitro* and *in vivo* experiments. Nevertheless investigations on most saprophytic filamentous fungi are rare and fragmentary. The problem caused by fungi is not restricted to the realm of human and animal health. It is rather multifaceted, affecting different types of industry products such as paper and leather, among several others. The antimycotic property of propolis may thus be utilized in preventing the destruction of books and other valuable paper products in libraries and archives, and leather products by saprophytic fungi such as *Aspergillus sp., Penicillium sp.,* and *Trichoderma sp.*, which are generally known for rotting and destruction of such products. For this reason one of the aims of the present investigation was to test the antimycotic property of propolis *in vitro*.

The medical use of propolis was nearly forgotten in modern era due to the discovery and effective use of antibiotics. Nowadays, however, since several pathogens are developing resistance to potent antibiotics, and the latter causing side effects in humans, the need to search and screen for new antimicrobial agents is increasing (Walker 1996, Mc Devitt et al. 2002). This gave an impetus to several researchers to concentrate on research in potentially antimicrobial natural products such as propolis to add to our current arsenal. Because propolis is reputed to have several biological effects against various types of ailments, the list of pharmaceutical, health food and cosmetic preparations is increasing drastically (Burdock 1998).

Regardless of the increasing emergence of drug resistant microbes, the pace at which new antimicrobials are discovered and produced is slowing and the so called new and emerging pathogens are aggravating the problem (Russell 2002). The mechanisms of antimicrobial actions of antibiotics, and the resistance mechanisms by most microbes to antibiotics are well documented (Russell and Chopra 1996, Nikaido 1998). The mechanisms of action of biocides based on natural mixtures, such as propolis, are however, poorly understood due mainly to the several target sites they have within a bacterial cell (Denyer and Stewart 1998), and research to elucidate these mechanisms is inadequate. Therefore screening methods that contribute to the understanding of the mechanisms how propolis act are very important. However, as the need to isolate potent antimicrobial agents is increasing, several techniques that may not give a clue as to the mechanism of action of the biocide are employed in several laboratories. In addition to that, most standard antimicrobial screening methods are associated with drawbacks, and there is no standard way of presenting the results obtained with these methods (Rios et al. 1988). Few of the most common problems encountered with the majority of techniques employed today are: (i) different inoculum sizes used in various laboratories: this can affect results in both liquid and solid media directly, and different results could be obtained for the same concentration of an antimicrobial agent based on the density of microbes to be treated. This problem could be minimized to a certain degree by standardizing the inoculum density which is, however, laborious, and time consuming, (ii) consistency of the agar layer (activity of water): affects results by directly interfering with the diffusion potential of the compound being tested, (iii) incubation temperature: affects the diffusion potential of the substance being screened by acting on the consistency of the agar layer, (iv) polarity of the antimicrobial substance being tested: testing non-polar substances (like propolis) with the Petridish bioassay method is not as such credible since the substance could not diffuse properly through the polar agar layer, and hence antimicrobial activity could be highly restrained and restricted to a small diameter around the hole containing propolis solution, (v) colour of the substance to be tested: this interferes with results from the spectrophotometric method. It could, however, be alleviated to a certain extent by using corresponding blanks for each dilution, but it can not be reliable. The problem with propolis is that, in addition to its natural colour which could interfere with absorbance by the bacterial cells, it forms a hazy suspension upon mixing with water due to the insolubility of most of its components in water that makes spectrophotometric measurement impractical.

A method which can not be affected by these problems is of interest in the search for antimicrobial agents. One of such methods that is robust enough to be used in the study of antimicrobial agents is calorimetry. This technique is applied in different fields of science with high precision and sensitivity (Lamprecht 1983). One of the aims of the present investigation is, therefore, to use calorimetry in the investigation of the mechanism of action of propolis and evaluate its credibility compared to the often-used standard microbiological methods in testing the effects of antimicrobials.

Research has been done on the biological activity of propolis against different sorts of ailments, infections and parasites in the past. But most investigations concentrated on a sample from 1 geographic location, 1 type of extract (usually the ethanol extract of propolis, EEP), or derivatives of 1 type of extract, and almost all used only the Petridish bioassay method with no, or very little hints about the mechanisms behind the antimicrobial effects. One of the purposes of the present investigations was to compare differences among the antimicrobial activities of propolis extracts from different geographic origins, and samples from the same apiary but different hives. The comparison between propolis samples requires considering the species and subspecies of bees that did the collection. The comparison between geographic samples will be done at the level of different extracts, i.e. ethanol-extracted propolis (EEP), water-extracted propolis (WEP), and propolis volatiles (PV).

7.3 Materials and Methods

7.3.1 Propolis acquisition and preparation of ethanol extracted propolis (EEP)

Propolis samples were obtained from different parts of the world by personal contact with beekeepers and scientists in the corresponding countries. The different propolis samples used in the investigations were tabulated in chapter 5 with their physical properties and the bees that collected them. According to the information obtained from the corresponding propolis suppliers, all propolis samples were collected by scrapping from frames and walls of the beehives. All samples were obtained as solid samples and extracted in 70% ethanol according to previously established and effective methods (see chapter 3). The extracts were cooled to 25 °C, and their electrical conductivity and pH were measured using a multichannel

WTW Multi 340i conductometer and pH meter (Wissenschaftlich-Technische Werkstätten GmbH, Weilheim, Germany). The propolis extracts were dried at 40 °C by monitoring the weight at an interval of 24 h starting from the 10th incubation day. Complete dryness of the samples was ascertained by the absence of weight loss during the last 3 continuous weightings, and was achieved in about 15 days. The yield of extraction of each propolis sample was determined from the proportion of dry weight of extracted sample to that of fresh sample. The pH of the different propolis samples and yield of extraction are summarized in Table 7.1.

The dried propolis samples were dissolved in 60% ethanol to prepare 10% stock solutions, which were diluted further with the same solvent to achieve varying experimental concentrations. Though extraction of propolis was done with 70% ethanol, treatments were done with 60% ethanol to minimize the effect of a highly concentrated solvent.

7.3.2 Extraction of propolis volatiles (PV)

As it is often claimed that the volatile components of propolis are responsible for the lower density of aeroflora in the otherwise ideal bacterial and fungal flourishing beehive interior, the antimicrobial activities of the volatile components were investigated after extracting them. Due to limitation of sample size, antimicrobial activities of the volatile components of propolis (propolis volatiles, PV) were investigated only with samples from Colombia (C1), Ethiopia (E1), Germany (G1), Italy (I1), 2 samples from Poland (P1 and P2) and 3 samples from South Africa (SA1, SA3, and SA5).

The pre-weighed propolis samples were frozen at -20 °C to make handling of this highly sticky substance easier. The samples were then ground in an electric coffee mill and subjected to steam distillation for 4 h using a Lickens-Nickerson apparatus (Kujumjiev et al. 1999). The collected distillates were extracted with ethyl ether/n-pentane 1:1, and the extracts were dried over Na_2SO_4. The dried samples were weighed and dissolved in 60% ethanol to make a stock solution of 10% w/v (10 g/100 ml) for further use.

7.3.3 Water-extracted propolis (WEP)

Due to limitations of sample sizes obtained from different countries, it was not possible to prepare a water-extracted propolis solution of all samples. Such a solution was prepared only for the sample obtained from Berlin, Germany (G1). As the yield of extraction is very low compared to the ethanol-extracted propolis (see results), a large sample size was needed to get a reasonable amount of extract to perform experiments.

The extraction procedure used to obtain the ethanol-extracted propolis (section 7.2.1 above) was followed strictly except that the solvent here was distilled water instead of 70% ethanol. A 10% stock solution was prepared in distilled water and lower concentrations for treatments were prepared by diluting the stock solution with distilled water.

7.3.4 Biological material

Bioassays of the antimicrobial activities of the different propolis samples were performed using (i) 4 fungal species: the yeast *Saccharomyces cerevisiae* (DSM 211) and 3 filamentous fungi, *Aspergillus niger* (DSM 737), *Penicillium chrysogenum* (DSM 844) and *Trichoderma viride* (DSM 63065); (ii) 4 species of Gram positive bacteria: *Bacillus subtilis* (DSM 347), *Micrococcus luteus* (DSM 348), *Bacillus megaterium* (DSM 90), *Bacillus brevis* (DSM 5609); and (iii) 2 species of Gram negative bacteria: *Escherichia coli* (DSM 31), *Pseudomonas syringae* (DSM 5176). All strains of microorganisms were bought from the German collection for microorganisms and cell culture (Deutsche Sammlung von Mikroorganismen und Zellkulturen GmbH - DSMZ, Braunschweig).

7.3.5 Growth media

Cultivation of all bacterial cultures was done in Standard I nutrient broth (Merck Lot VL 630582) and/or on Standard I nutrient agar (Merck, Lot VL 694681). The yeast was cultivated on a medium composed of 10 g Oxoid agar (Agar Bacteriological No.1, Lot 817706-2), 1 g yeast extract (Sigma, CAS No. 8013-01-2), 2 g glucose (Merck), 0.5 g peptone (Sigma, Lot 128H0184), 10 ml Na-phosphate buffer (1M, pH 7.0) in 1 l distilled water. The same medium without agar was used as nutrient broth for the yeast.

Trichoderma viride was cultivated on malt extract peptone agar (MEPA) composed of 30 g malt extract (Sigma, Lot 41k0181), 3 g soya peptone (Hy soy T, Sigma, Lot 128 H0184), 15 g Oxoid agar (Agar Bacteriological No.1, Lot 817706-2) in 1 l distilled water. *Aspergillus niger* and *Penicillium chrysogenum* were cultivated on potato dextrose agar (PDA) composed of 20 g glucose (Merck), 15 g Oxoid agar (Agar Bacteriological No.1, Lot 817706-2), in 1 l potato infusion obtained by boiling 200 g scrubbed and sliced potato in 1 l distilled water for 1 h and passed through a fine sieve, according to the recommendation of the DSMZ obtained up on purchase of the strains.

All types of media were sterilized by autoclaving at 121 °C and 1.03 bars for 15 min.

7.3.6. Petridish Bioassay

7.3.6.1 Agar well diffusion pour plate technique

The main purpose of this experiment was to compare the antimicrobial effects of the various propolis samples at levels of different concentrations. As the general procedures for inocula preparation and treatment are different, bacteria and yeasts are dealt in here separately from the filamentous fungi.

Propolis concentrations of 0.01, 0.10, 1.0, and 10.0% w/v were prepared by diluting the stock solution with 60% ethanol in case of EEP and with distilled water for WEP, and used in the antimicrobial tests. In the case of bioassays with propolis volatiles, experiments were done only with a 10% concentration for all samples due to sample size limitation.

Except for the WEP, where distilled water was used, the control for all experiments was 60% ethanol. Each experiment was repeated 3 times.

7.3.6.1.1 Bacteria and yeast

An isolated pure colony of an overnight grown culture was picked carefully using a sterile transfer loop, inoculated into a nutrient broth in an Erlenmeyer flask and grown overnight at 30 °C. A volume of 50 µl of the overnight culture was inoculated to a 20 ml nutrient broth and grown further for about 3 to 5 h until an O.D. of 0.6 (546 nm) was achieved. Optical density of the culture was measured using a spectrophotometer of the type UV 120-01 Shimadzu, Kyoto, Japan. The suspension was then diluted 1:50 with the corresponding nutrient broth in order to prepare the standard inoculum (method slightly modified after Faye and Wyatt, 1980).

The sterilized nutrient agar was cooled to 48 °C, 5 ml of the standard inoculum was mixed with 1 l nutrient agar and distributed into plastic Petridishes of ø = 85 mm, 10 ml in each, rendering an agar layer of 3.52 mm thickness. After the agar was solidified, 3 holes were bored per Petridish using a cork borer of ø = 9 mm. Each hole was then filled with 50 µl propolis solution and the Petridishes were placed in a refrigerator for 24 h, giving propolis sufficient time to diffuse. Finally, the plates were removed from the refrigerator, incubated at 30 °C for 24 to 48 h and the inhibition zones were measured. Antibacterial tests were done using propolis concentrations of 0.01, 0.1, 1.0, and 10.0 (% w/v).

7.3.6.1.2 Filamentous fungi

An isolated pure colony of a fungal culture, which was grown for 72 h on solid medium and started to sporulate, was scrubbed up using a sterile transfer loop and put into sterile distilled water in a test tube. The hyphae were disintegrated by adding sterile glass beads and shaking vigorously for 1 min, in order to get a uniform suspension. The optical density of the suspension was measured at 546 nm and adjusted to 0.6. The suspension was then diluted 1:50 with sterile distilled water in order to prepare the standard inoculum. 5 ml suspension (standard inoculum) was added to a 1 l agar solution at 48 °C and the same procedure as above (section 7.3.6.1.1) was followed further until the final incubation phase. The plates were taken out of the refrigerator and incubated at 25 °C for 72 h and the inhibition zones were measured. Propolis concentrations of 0.1, 0.5, 1.0, 2.0, 4.0, 8.0, and 10.0 (% w/v) were tested, except for propolis volatiles with only 10%.

7.3.6.2 Agar dilution in plates

As the antimicrobial activity of propolis by the agar well diffusion pour plate method is highly influenced by the low hydro-solubility of biologically active components in the diffusion medium (the agar layer) estimation of the MIC values using this method may not be accurate (Sawaya et al. 2002). For this reason MIC values for the different propolis samples against the various bacterial and fungal species were determined by the agar dilution method, according to the recommendation of the National Committee for Clinical Laboratory Standards guidelines (NCCLS, 1985).

Corresponding volumes of a 10% propolis solution or of lower concentrations were added to the sterile agar solutions at a temperature of 48 °C, to achieve final concentrations of 0.05, 0.1, 0.5, 1.0 1.5, and 2% w/v propolis in the fungal growth media and 0.005, 0.01, 0.02, 0.04, 0.06, 0.08, 0.1% w/v propolis in the bacterial and yeast growth media. In addition to that final propolis concentrations of 1, 5 and 10% w/v were incorporated in the media for bacterial strains that were insensitive to the lower concentrations. The contents were mixed thoroughly, poured into sterile Petridishes, and allowed to cool. The same concentrations of propolis volatiles, as above, were tested against bacterial and fungal strains for their minimal inhibitory concentrations. Standard inocula were prepared for all bacteria, yeast and fungi following the same procedure as in the agar well diffusion pour plate method. Inoculation was done by pipetting 50 µl of the standard inoculum and uniformly distributing it on the surface of the propolis-containing agar layer using a sterile z-shaped glass rod. The plates were then incubated at a temperature of 30 °C for 24 h for bacteria and at 25 °C for 48 to 72 h for fungi. The minimal inhibitory concentration (MIC) was the lowest concentration of propolis that inhibited any visible growth of bacteria, yeast, or fungi.

7.3.7 Calorimetric Bioassay

All calorimetric experiments were performed with bacteria at a temperature of 30 °C using a flow calorimeter (Type 10700-1, LKB Bromma, Sweden) with a flow-through spiral of 0.587 ml. The calorimeter was connected by a Teflon tube of 1 mm inner diameter to an external fermenter, a 50 ml reaction vessel with 20 ml nutrient broth, placed in a water bath at 30 °C. The bacterial culture was circulated from the fermenter to the calorimeter and back using a peristaltic pump (Type LKB Pharmacia, Bromma, Sweden) at the outlet of the calorimeter in a sucking mode. The culture was vigorously stirred with a magnetic stirrer in order to avoid settling of cells and minimize depletion of oxygen in the fermenter and in the flow line.

7.3.7.1 Calibration of the flow calorimeter

As the sensitivity of the calorimeter and hence results could be affected, among several other factors, by the pumping rate, the calorimeter was electrically calibrated by pumping a phosphate buffer of pH 7 at pumping rates ranging from 1.5 to 100 ml h^{-1} and an experimental temperature of 30 °C. The sensitivity of the flow calorimeter decreases linearly with increasing pumping rate (see appendix Fig. A1).

To satisfy the contradicting requirements for a high flow rate (to avoid settling of cells and exhaustion of oxygen during the residence time in the tubing system) and a low one (to allow sufficient temperature equilibration of the culture before it arrives in the calorimetric spiral) a pumping rate of 56 ml h^{-1} was used for investigations with a prevailing aerobic metabolism. The sensitivity of the calorimeter at this pumping rate was 61.6 μV mW^{-1}.

7.3.7.2 Sterilization of the calorimetric set-up

The flow calorimetric line and the calorimetric spiral were sterilized by circulating a sterilizing solution composed of 10% H_2O_2 and 2% H_2SO_4 in 60% ethanol for 30 min before and after each experiment. After the allocated sterilization time the flow calorimetric set up was cleaned with 0.1 M potassium-phosphate buffer of pH 7.0 for 1 h. This sterilizing solution, in addition to sterilizing the calorimetric set up, washes out cells attached to the wall, and cleans any water insoluble propolis residue that could precipitate and settle on the inner wall of the tubes or the calorimetric spiral and subsequently block the flow system. Blockage of the flow-line by the settling of the water insoluble components of propolis was a very common phenomenon at low pumping rates and high cell densities after treatment with propolis. In order to avoid this problem a higher pumping rate, which reduces settling due to the higher rate of streaming of fluid during the experiment, and the relatively aggressive cleaning and sterilizing solution that could dissolve and remove any precipitate from the tubing system were used.

7.3.7.3 Microcalorimetric cultivation of bacteria

As cultivation of moulds in a flow calorimetric setup is impossible due to the impracticality of pumping a broth containing a hyphal suspension, these calorimetric investigations were done only with bacteria. Even though it is in principle possible to cultivate yeasts in the flow calorimetric setup, like in the case of bacteria, it is very difficult and time consuming to deal with yeast cultures as they intend to cling to the inner wall of the tubing and block the flow system. For this reason flow calorimetric experiments with the yeast culture were avoided.

As recording of only the heat production rate may not explain what is happening in the tubing system and in the fermenter, oxygen consumption rate and the number of colony forming units (CFU) were recorded simultaneously for untreated, control bacterial cultures of all strains.

After the flow line was properly sterilized and cleaned with buffer, 20 ml of the growth medium (Standard I nutrient broth) were circulated for at least 30 min to establish a stable baseline. An inoculum of 200 µl of an overnight culture, grown on Standard I nutrient broth, of the experimental bacteria was then added and allowed to grow. Preliminary calorimetric experiments coupled with polarography and CFU count displayed that all strict aerobes, i.e. *B. megaterium, B. subtillis, B. brevis, M. luteus,* and *P. syringae*, have similar patterns of *p-t* curves, change in oxygen concentration in the flow line and of number of CFU, though minor differences exist among the shape of the *p-t* curves. The facultative anaerobe *E. coli*, however, showed a unique *p-t* curve due to the shift of metabolism to the anaerobic phase. For this reason, further calorimetric experiments were done using *E. coli* as a typical facultative anaerobe and *B. megaterium*, randomly chosen as a representative strict aerobe. The calorimetric signals were amplified (1000x) and recorded as power-time (*p-t*) curves by a two-channel recorder (BD5, Kipp and Zonen, The Netherlands). The recorded curves were digitalized using a Planimeter (Digikon DK 4261, Kontron Registriertechnik GmbH, München, Germany) and directly transferred to a PC. The ASCII data were then imported to statistic and graphic PC programmes for further processing. The results of the calorimetric experiments were presented as volume specific heat production rates (*p* in $\mu W\ ml^{-1}$) as a function of time.

7.3.7.4 Determination of O_2 concentration and CFU

The tube connection between the fermenter and the calorimeter, though it has to be kept as short as possible to avoid exhaustion of oxygen and hence the possible physiological change of the organism, is unavoidable. Though it is undesirable due to the mentioned facts, it however, provides space for the incorporation of electrodes in the flow line for the simultaneous monitoring of other culture characteristics, in addition to the heat production rate.

It is often mentioned that the concentration of oxygen decreases with increasing cell density, and becomes a limiting factor for bacterial growth and metabolism in the tubing system of the flow calorimeter. For this reason the concentration of oxygen was monitored by incorporating 2 galvanic oxygen electrodes, 1 in the fermenter, and the other in the flow line at the outlet of the calorimeter. The oxygen electrodes used were both of the type WTW Cellox 325, connected to WTW Multi 340i Data logger (Weilheim, Germany).

As the sterilizing chemicals were too aggressive to be used with the membrane oxygen electrodes, the latter had to be sterilized separately with only 70% ethanol. Due to the labour intensiveness of the procedure of dismantling the setup, sterilizing the electrode separately and assembling the flow-line with the oxygen electrode incorporated, this experiment was done only in few cases as a control. In addition to its labour intensiveness, the procedure of sterilizing the electrodes separately and assembling finally, increases the chance of contaminating the sterile flow line.

The number of CFU was determined by removing 50 µl culture from the outlet of the calorimeter (inlet of the fermenter) every 30 min. The samples were serially diluted and plated on Standard I nutrient agar, incubated for 24 h at 30 °C and the number of CFU was counted. Plates containing CFU < 30 were rejected in order to avoid possible experimental error.

7.3.7.5 Treatment of bacteria with propolis

Due to facts that will be dealt with in the discussion part of this chapter, it was decided to do the treatments only in the exponential phase of bacterial growth. In addition to that the effect of treatment at this phase is clearly visible by the drop in the level, or the change in the slope of the *p-t* curve, which is directly related to the growth/metabolic rate of the bacterial culture, and can be compared with other treatments. Treatment of bacteria with propolis was done in the exponential growth phase by adding 10, 25, 50 or 100 µl of the 10% ethanol-extracted or volatile components of propolis stock solution to the fermenter with 20 ml bacterial culture, to achieve final propolis concentrations of 0.005, 0.0125, 0.025 or 0.05% (w/v). As the water-extracted propolis solutions were ineffective at these concentration levels, more volumes of the stock solution (10x) were added to the culture to achieve concentrations of 0.05, 0.125, 0.25 or 0.5% (w/v). The experiments with water-extracted and volatile components of propolis were done only with the sample G1 due to its sufficient availability.

In addition to the treatments at the exponential growth phase some experiments were also done by treating the culture with moderate concentrations of propolis after the calorimetric curve had dropped to the "calorimetric death phase".

The control experiments for each concentration of propolis were done by adding an equal volume of 60% ethanol, to observe the antimicrobial activity contributed by the solvent. Distilled water, instead of ethanol, was used as a double control in order to observe the influence on the heat production rate, if any, of the "dilution effect" of the liquid added rather than antimicrobial activity.

7.3.7.6 Determination of calorimetric MIC and MBC values

The minimum concentration of propolis that inhibited visible bacterial growth after incubation for a given period in a Petridish assay experiment was considered as the MIC value against the corresponding bacterial strains. In case of the calorimetric experiments, however, the minimum concentration of propolis that resulted in a drop of the p-t curve, was considered as the MIC value against the corresponding bacteria.

The minimum bactericidal concentrations (MBC) of the different propolis samples were also determined calorimetrically. The minimal concentration of propolis that killed bacteria and hence dropped the heat production rate to the baseline either immediately or first to a level above the base line and gradually, with incubation time, to the baseline were considered as the MBC.

7.3.8 Statistical analysis

Results were presented as mean ± S.D. values. Statistical tests were performed using the two-tailed student's t-test, paired sample t-test, 2-way ANOVA, 3-way ANOVA, and Tukey's post-hoc test, according to the nature of the data, and $\alpha = 0.05$ was taken as the critical value for all tests.

7.4 Results

7.4.1 Differences in the yields of extraction and physical properties of propolis samples

The pH of extracted propolis samples was acidic between 4.2 and 5.3, showing slight differences among different geographic samples, samples from the same geographic region and even from the same apiary. The pH was independent of the method of extraction as the 3 different types of extracts, i.e. EEP, WEP, and PV had similar pH values. The yields of the 3 extraction methods were highly variable; steam distillation extracted only the volatile components of propolis, and yielded 0.2% to 0.9% w/w, water extraction yielded 3.5% w/w, whereas the ethanol extraction method showed a very high yield of up to 61.3% w/w. There was also a high variation in the yield of extraction of the different propolis samples using the ethanol extraction method, with yields ranging from 10% for SA3 to 61.3% for P2. Variation was also observed in the yield of propolis volatiles (PV), which varied from 0.2% for SA1 and SA3 to 0.9% for G1. As the extraction of the water soluble components of propolis was done for only 1 sample, comparison between samples could not be done at this point. The yield of propolis showed dependency on the quality of raw propolis. Those propolis samples that were pure and sticky when they were still raw had higher yields. However, even though C1 and P4 had impurities, they demonstrated higher yields; whereas even though K1 was pure propolis it had a lower yield compared to the other pure propolis samples (Table 7.1).

Table 7.1 Physical properties and yield of extraction of the propolis samples
Yield (% w/w) of EEP (ethanol-extracted propolis), of WEP (Water-extracted propolis), and of PV (propolis volatiles). Electrical conductivity (C) in $\mu S\ cm^{-1}$, pH value, and description of the raw propolis quality. (The conductivity values are after correction for the corresponding solvents since distilled water and 60% ethanol have different conductivities, as ethanol is a non electrolyte.)

Sample name	EEP			PV			Remarks
	pH	C	yield	pH	C	yield	
C1	4.0	22.3	35.7	4.1	15.3	0.6	Impure propolis, slightly sticky
E1	5.3	25.4	44.7	5.0	12.4	0.8	Pure propolis, very sticky
G1	4.4	21.2	56.2	4.4	16.3	0.9	Pure propolis, very sticky
K1	4.8	33.4	31.2	-			Pure propolis, very sticky
Rus1	4.9	16.7	55.0	-			Pure propolis, very sticky
P1	4.8	44.2	54.5	4.9	22.3	0.5	Pure propolis, very sticky
P2	4.7	26.6	61.3	4.6	11.6	0.6	Pure propolis, slightly sticky
P3	4.7	24.2	59.3				Pure propolis, very sticky
P4	4.9	15.4	47.9				Impure propolis, slightly sticky
SA1	5.2	51.2	16.4	5.0	25.8	0.2	Impure propolis, slightly sticky
SA3	5.1	22.6	10.6	5.3	10.3	0.2	Impure propolis, slightly sticky
SA5	4.9	29.3	22.6	4.6	11.2	0.3	Impure propolis, slightly sticky
SA6	4.5	17.4	24.8				Impure propolis, slightly sticky
SA8	5.0	42.3	22.7				Impure propolis, slightly sticky
SA11	4.6	33.6	13.1				Impure propolis, slightly sticky
I1	4.9	36.5	55.3				Pure propolis, very sticky
WEP	4.3	98.9	3.5				Pure propolis, very sticky

7.4.2 Antimicrobial activities of ethanol extract of different propolis samples

Comparisons of the antimicrobial activities of the various propolis samples against the different test organisms were done at several levels of propolis concentrations using the diameter of inhibition zone as a parameter. A 3-way analysis of variance ANOVA ($\alpha = 0.05$) followed by the Tukey's post-hoc test using the variables: bacterial strain, propolis type, and concentration of propolis displayed that significant differences exist ($p < 0.05$) between the antibacterial activities of different concentrations within a sample, the activity increasing with increasing concentrations above the MIC values for each. However, this trend was not always observed. No statistically significant difference existed ($p > 0.05$) in some cases in the antifungal activities of the different concentration of a propolis sample, i.e. the diameter of the inhibition zone remained almost constant as the concentration changed stepwise from 1% to 10% (see appendix, Table A2).

Considering the antibacterial activity of the various propolis samples against a bacterium at a certain concentration level, it becomes clear that the samples showed antibacterial activities slightly different from each other, but no statistically significant difference, regardless of their geographic origins (see appendix, Table A1).

The slight differences observed in the antimicrobial activities of the different propolis samples could also not be explained based on the species of bees that collected them, since the difference within a species could be as high as or even higher than that between species. The samples SA1 to SA11 were collected by *Apis mellifera capensis* (South Africa), E1 by *A. mellifera scutellata* (Ethiopia), P1 to P4 by *A. m. mellifera* (Poland), G1 by *A. m. carnica* (Germany), I1 by *A. mellifera ligustica* (Italy), C1 by *Tetragonisca angustula* Illger (stingless bee) (Colombia).

7.4.3 Comparison of antimicrobial activities of different extracts

Comparison of the 3 different groups of extracts, i.e. ethanol-extracted propolis (EEP), water-extracted propolis (WEP) and the propolis volatiles (PV), was done by considering the minimal inhibitory concentrations (MIC) (Tables 7.2, 7.3, 7.4) and the inhibition diameter of a 10% propolis sample (see appendix Tables A1, A2 and A3). All propolis samples were extracted with 70% ethanol and thus had an ethanol-extracted propolis component. But due to sample size limitation, PV extracts were prepared only for C1, G1, E1, I1, P1, P2, SA1, SA3 and SA11 and WEP was obtained only from G1.

Table 7.2 Minimal inhibitory concentrations (MIC) of propolis against bacteria
MIC values (% w/v) of EEP from different geographic origins, and of a WEP, against bacterial species, determined by the agar dilution on plate method. n.d. indicates that no inhibitory concentration was detected in the range tested, up to 10% w/v.

Propolis type	*B. brevis*	*B. megaterium*	*B. subtilis*	*M. luteus*	*E. coli*	*P. syringae*
WEP	0.500	10.000	10.000	10.000	n.d.	10.000
I1	0.010	0.010	0.060	0.100	0.500	0.060
E1	0.005	0.060	0.040	5.000	n.d.	0.080
C1	0.040	0.040	0.040	0.500	n.d.	0.080
K1	0.005	0.010	0.080	0.060	0.100	0.060
G1	0.010	0.010	0.040	0.040	0.100	0.500
RUS1	0.010	0.010	0.020	0.020	5.000	0.040
P1	0.010	0.010	0.040	0.080	0.100	0.040
P2	0.010	0.010	0.040	0.500	0.100	0.500
P3	0.040	0.010	0.040	0.060	0.100	0.100
P4	0.060	0.010	0.040	0.060	0.100	0.100
SA1	0.005	0.010	0.005	0.020	0.100	0.005
SA3	0.005	0.060	0.005	0.040	5.000	0.060
SA5	0.005	0.005	0.005	0.005	1.000	0.005
SA6	0.005	0.005	0.005	0.005	5.000	n.d.
SA8	0.005	0.005	0.005	0.060	1.000	0.010
SA11	0.005	0.060	0.060	0.080	1.000	0.040

Table 7.3 Minimal inhibitory concentrations (MIC) of propolis against fungi
MIC values (% w/v) of EEP from different geographic origins, and of a WEP against various filamentous fungi and a yeast determined by the agar dilution on plate method. n.d. indicates that no inhibitory concentration was detected in the range tested, up to 10% w/v.

Propolis type	*T. viridae*	*A. niger*	*P. chrysogenum*	*S. cerevisae*
WEP	n.d.	n.d.	n.d.	5.00
I1	1.00	1.00	1.00	0.10
E1	n.d.	2.50	1.00	0.50
C1	n.d.	0.50	1.50	0.10
K1	0.50	0.50	0.50	0.04
G1	1.50	1.50	1.50	0.10
RUS1	2.00	1.00	1.50	0.50
P1	1.00	1.50	1.00	0.04
P2	1.00	1.00	0.50	0.50
P3	0.50	1.00	0.50	0.04
P4	0.50	1.00	0.50	0.10
SA1	n.d.	1.00	1.00	0.10
SA3	n.d.	1.00	1.00	0.50
SA5	10.00	1.00	1.00	0.06
SA6	n.d.	1.00	1.00	0.08
SA8	2.00	1.00	1.00	0.01
SA11	2.00	1.00	1.50	0.50

The water-extracted propolis (WEP) was shown to be significantly less active ($p < 0.05$, t-test) than the ethanol-extracted one from the same apiary (G1) as displayed by the higher MIC value and also by the smaller diameter of inhibition zone achieved by the 10%

WEP against each organism tested. Inferiority of the antimicrobial action of WEP of G1 also holds true when compared to the ethanol-extracted propolis samples obtained from different geographic regions.

Table 7.4 Minimal inhibitory concentrations (MIC) of propolis volatiles (PV)
MIC values (% w/v) of propolis from different geographic origins against various bacterial and fungal species determined by the agar dilution on plate method. n.d. indicates that no inhibitory concentration was detected in the range tested, up to 10% w/v.

	E1	C1	G1	P1	P2	SA1	SA3	SA5	I1
B. brevis	0.01	0.10	0.08	0.04	0.04	0.01	0.01	0.01	0.08
B. megaterium	0.10	0.10	0.08	0.04	0.04	0.04	0.10	0.04	0.08
B. subtilis	0.08	0.10	0.08	0.08	0.08	0.01	0.01	0.01	0.08
M. luteus	n.d.	1.00	0.08	0.10	1.00	0.06	0.08	0.02	0.08
E. coli	n.d.	n.d.	0.50	0.50	0.50	0.50	n.d.	5.00	0.50
P. syringae	0.10	0.10	1.00	0.06	1.00	0.01	0.01	0.01	0.50
S. cerevisae	1.00	0.50	0.50	0.08	1.00	0.50	1.00	0.10	0.50
A. niger	5.00	1.00	5.00	2.50	2.50	5.00	5.00	5.00	2.50
P. chrysogenum	5.00	2.50	5.00	2.50	2.50	5.00	5.00	5.00	2.50
T. viridae	n.d.	n.d	5.00	2.50	2.50	n.d.	n.d	n.d	n.d.

A 10% concentration of the volatile components of propolis (PV) showed slightly, but not statistically significant ($p > 0.05$, t-test) weaker antimicrobial activities than the ethanol-extracted one, against each microbe (cf. the 10% inhibition columns of Table A1 and A2 with that of Table A3 in the appendix). The t-test for statistical significance was performed for each pair of extracts i.e., EEP versus PV of each propolis sample against every bacterial and fungal species tested. All PVs showed significantly higher antimicrobial activities compared to the WEP of the sample G1.

The minimal concentrations of propolis needed to inhibit microbial growth were higher in case of the PVs than the EEPs. A 2 to 10 fold concentrated PV was needed in order to get a complete inhibition of bacterial growth as would be achieved by the EEP of the same propolis sample (cf. Tables 7.2 and 7.3 with Table 7.4). The filamentous fungi were less sensitive or even insensitive to the volatiles components of propolis at lower concentrations, as in the case of the ethanol extracts of propolis. The sensitivity of bacteria and fungi demonstrated by the diameter of the inhibition zone was not clearly segregated at a higher PV concentration (10% w/v), except for *B. brevis* which showed significantly higher inhibition zone for all PV samples tested. Bacteria that were sensitive to only highly concentrated ethanol extracts of propolis (*M. luteus* to 5% E1 and *E. coli* to 5% SA3) were insensitive even to a 10% PV.

7.4.4 Differences in the sensitivity of various microorganisms to propolis

Comparison of sensitivity of the different test organisms in view of the MIC values displayed in Tables 7.3 and 7.4, and also of the inhibition diameters displayed by higher concentrations (see appendix Table A1 and A2), demonstrate that filamentous fungi are generally less sensitive to propolis treatment. The minimal inhibitory concentrations of the various propolis samples against the filamentous fungi tested lies between 0.5 and 2.5% w/v propolis (Table 7.3), and 0.005 to 0.5% w/v against the different bacteria except *E. coli* (Table 7.2).

The yeast showed significantly higher MIC values than most bacteria, but significantly lower ones than the 3 moulds. Nevertheless, at 10% w/v the inhibition zones were similar to that obtained for most bacteria. The 4 Polish (P1 to P4) and the Kazakh (K1) propolis samples showed very large inhibition zones against the yeast *S. cerevisae*, significantly higher than that achieved for any other organism tested.

Among the bacteria, the Gram negative bacterium *E. coli* was highly resistant to propolis treatments followed by the other Gram negative *P. syringae* (see appendix, Table A1). *E. coli* did not show any recognizable response to the 10% treatments with E1, C1 and WEP, and small inhibition diameters with the other propolis samples. Only the Russian and Kazakh propolis produced larger inhibition diameters (see appendix, Table A1).

In the agar well diffusion pour plate method, treatments with propolis concentrations lower than the MIC values stimulated dense growth of *P. syringae* around the hole containing propolis (Fig. 7.1). Concentrations slightly higher than the MIC values produced a small inhibition zone around the hole, usually surrounded by a zone of dense bacterial growth (Fig 7.3 a and b, and see appendix Table A1). No encouragement of bacterial growth was observed at higher concentrations of propolis where only inhibition zones of different diameters were observed. This phenomenon of encouragement of bacterial growth at concentrations around the MIC values was not observed for any other organism tested, it was a unique feature observed for *P. syringae*.

The 2 moulds *A. niger* and *P. chrysogenum* showed similar sensitivities to the treatments with all propolis samples. However, *T. viridae* was insensitive to 3 of the 6 South African samples (SA1, SA3 and SA5), to the sample from Ethiopia (E1), from Colombia (C1) and the water-extracted propolis (WEP) at a 10% concentration. All filamentous fungi were insensitive to 10% WEP

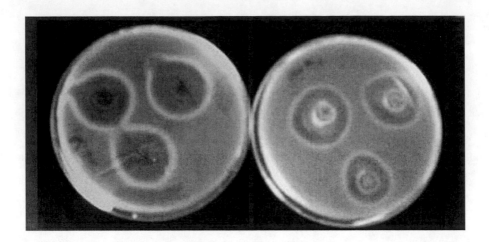

Fig. 7.1 Hormesis effect of a 0.1% E1 (left) and a 0.01% SA1 (right) EEP on *P. syringae*. The grey spots at the centre of the inhibition zones (in the well) are propolis precipitates, which did not diffuse due to hydrophobicity.

7.4.5 Calorimetric experiments
7.4.5.1 Calorimetric cultivation of bacteria

Since the patterns of curves of the heat production rates, oxygen concentration in the flow line and in the fermenter, and the number of colony forming units (CFU) were similar for all strict aerobes, treatments were done only with *B. megaterium* as a representative of them and with the facultative anaerobic *E. coli,* with a unique *p-t* curve. For this reason further calorimetric discussions and illustrations will mainly concentrate on these 2 bacterial strains. Even though the *p-t* curves of the strict aerobes investigated have similar patterns, they are not actually exactly alike since differences exist in the level at which the peak was achieved, 0.45 μW ml^{-1} for *B. megaterium*, 0.73 μW ml^{-1} for *B. brevis*, 0.52 μW ml^{-1} for *M. luteus*, 0.58 μW ml^{-1} for *B. subtillis*, and 0.57 μW ml^{-1} for *P. syringae*. In addition to that, differences also exist in the slope of ascend of the *p-t* curves, and slope of descend of the curve after the peak, and level of heat production rate at the calorimetric death phase. But the curves from the different strict aerobes have basic resemblances in shape, with room for minor differences in the features mentioned above due to specific features of each strain.

The microbial metabolic and growth events taking place in the fermenter during the first few hours of growth were represented by an initial lag phase followed by an exponential rise of the heat production rate and of the number of CFU. These events continued similarly up to the peak of heat production rate. After the *p-t* peak (0.45 μW ml^{-1}) for *B. megaterium*, the heat production rate dropped steeply whereas the number of CFU increased until to the stationary phase with a cell density of 1.47×10^8 CFU ml^{-1}, about twice as much as that at the peak (7×10^7 CFU ml^{-1}) about 2 h before. The heat production rate then remained at a lower level and the number of CFU at a higher level (Fig. 7.2 i). In case of *E. coli,* however, the nature of the *p-t* curve showed a different pattern. After the *p-t* curve achieved its peak at 0.62 µW ml^{-1}, it dropped down like in the case of the other bacteria but only to a level of about 0.21 µW ml^{-1}. After this point the rate of heat production rate increased and the curve started to ascend again until it achieved a level at about 0.5 µW ml^{-1}, lower than the first aerobic peak at 0.62 µW ml^{-1}(Fig. 7.2 ii) and remained at this level for the rest of the experimental period. It then started to drop gradually when the experimental period was extended for 10 to 15 h.

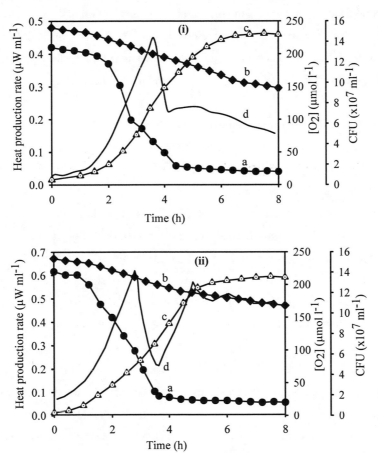

Fig. 7.2 Simultaneous recording of (a) oxygen concentration in the flow line, (b) oxygen concentration in the fermenter, (c) the number of colony forming units (CFU), and (d) the heat production rate of untreated cultures of **(i)** *B. megaterium* and **(ii)** *E. coli* in a flow microcalorimetric experiment.

The simultaneous monitoring of oxygen concentration in the flow line and in the fermenter displayed a big disparity between the two at higher cell densities in the middle and late experimental periods. At lower cell densities, at the lag phase and early exponential phase of bacterial growth, the concentration of oxygen in the flow line and in the fermenter were roughly similar, the latter showing a slightly higher concentration, by about 30 µmol l^{-1}. The beginning of the exponential growth was marked by an increase in the difference of the concentration of oxygen between that in the fermenter and in the flow line. The concentration of oxygen in the fermenter decreased gradually from 230 µmol l^{-1} at the lag phase to 152 µmol l^{-1} at the stationary phase. The concentration of oxygen in the flow line, however, dropped steeply from 200 µmol l^{-1} at the beginning of the exponential growth phase to 25 µmol l^{-1} shortly after the heat production peak was achieved. The peak of the calorimetric curve was achieved at an online oxygen concentration of 50 to 60 µmol l^{-1} and then declined with the steeply declining oxygen concentration for both *B. megaterium* and *E. coli*. The pattern of change of the oxygen concentration in the flow line and in the fermenter for *E. coli* and the other bacteria were alike regardless of the nature of the *p-t* curves (Fig 7.2 i and ii).

7.4.5.2 Effect of propolis treatment on bacterial culture properties

In order to evaluate the change in bacterial culture properties after treatment with lethal and sublethal doses of propolis, *B. megaterium* was treated with 0.025% and 0.05% w/v SA8, and *E. coli* with the same concentrations of P3.

After treatment of *B. megaterium* with 0.025% SA8, the heat production rate dropped suddenly from 0.39 to 0.15 µW ml^{-1} (61.5%) and the concentration of oxygen in the flow line and fermenter rose suddenly from 102 to 182 µmol l^{-1} and from 205 to 230 µmol l^{-1}, respectively. The number of CFU, however, dropped relatively slowly from 9.1 x 10^7 CFU ml^{-1} to 5.5 x 10^7 CFU ml^{-1} (Fig. 7.3 i). Following the drop in the heat production rate and CFU to lower values both stayed at plateau levels for nearly 2 h and then started increasing. The heat production rate increased further and achieved a peak at 0.42 µW ml^{-1} slightly lower than the peak of a control experiment (0.45µW ml^{-1}). The oxygen concentrations in the fermenter and in the flow line dropped with different rates, the one in the flow line dropping faster to a value of 25 µmol l^{-1} as the *p-t* curve dropped to a minimal value due to anaerobiosis. The online oxygen concentration, at which the *p-t* curve achieved its peak after treatment with the sublethal propolis dose, was 68 µmol l^{-1}, slightly but not significantly higher than the control experiments, 50 to 60 µmol l^{-1}.

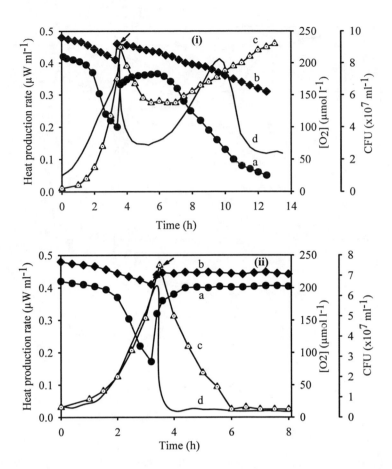

Fig. 7.3 Simultaneous recording of (a) oxygen concentration in the flow line, (b) oxygen concentration in the fermenter, (c) the number of colony forming units (CFU), and (d) heat production rate of a culture of *B. megaterium* treated with **(i)** 0.025% and **(ii)** 0.05% EEP of SA8 in a flow microcalorimetric experiment. Arrows indicate treatment.

After the *p-t* curve of *B. megaterium* dropped drastically due to the treatment of the culture with a lethal dose, it remained at about 0.03 μW ml^{-1} with a corresponding CFU of 4.5 x 10^6. Whereas the concentration of oxygen rose and remained at a higher level, at 198 μmol l^{-1} and 223 μmol l^{-1} in the flow line and in the fermenter, respectively (Fig. 7.3 ii). Unlike the changes in the rates of heat production and oxygen concentration in the flow line and in the fermenter, which are sudden, the number of colony forming units dropped gradually with incubation time, until it achieved its minimal level. *E. coli* responded similarly to the treatments with sublethal and lethal doses of propolis, as shown in Fig. 7.4 i and ii.

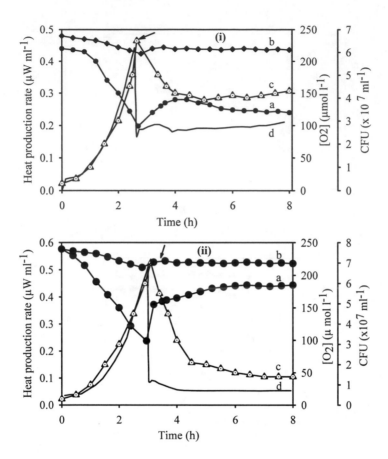

Fig. 7.4 Simultaneous recording of (a) oxygen concentration in the flow line, (b) oxygen concentration in the fermenter, (c) the number of colony forming units (CFU), and (d) heat production rate of a culture of *E. coli* treated with **(i)** 0.025% and **(ii)** 0.05% EEP of P3 in a flow microcalorimetric experiment. Arrows indicate treatment.

Treatment of *B. megaterium* with 0.025% EEP of SA8 after the *p-t* curve and the online oxygen concentration dropped to the minimal values, displayed that the 2 curves rose and attained values of 0.35 µW ml^{-1} and 147 µmol l^{-1}, respectively (Fig. 7.5). These increases were accompanied by corresponding increases in oxygen concentration in the fermenter and a slight decrease in the number of CFU. The curves then returned to the same level as before treatment in a short period of time.

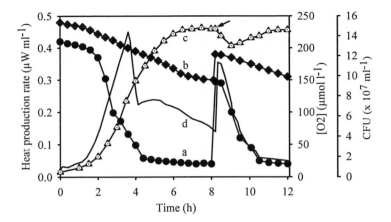

Fig. 7.5 Simultaneous recording of (a) oxygen concentration in the flow line, (b) oxygen concentration in the fermenter, (c) the number of colony forming units, and (d) the heat production rate of a culture of *B. megaterium* treated with 0.025% EEP of SA8 at the calorimetric death phase in a flow microcalorimetric experiment. The arrow indicates treatment.

7.4.5.3 Calorimetric dose-response curves

Comparison of the antibacterial activities of the different propolis samples were done calorimetrically by treating an exponentially growing culture of *B. megaterium* in a flow calorimeter with different concentrations of propolis from various geographic origins. The calorimetric curves after treatment of the culture with propolis (Fig. 7.6 i and ii) provided several information as to the kinetics of action of the different propolis samples. The following culture and curve after treatment-parameters were used in comparison with those before treatment, and/or of a control culture: (i) MIC and MBC, (ii) level of the *p-t* curve after treatment, (iii) slope of curve ascend after treatment, and (iv) peak of the *p-t* curve achieved after treatment.

7.4.5.3.1 Calorimetric MIC and MBC values

The minimum concentration of propolis that affected the *p-t* curve by dropping it to a level lower than that before treatment, regardless of the extent of drop, was considered as the minimal inhibitory concentration (MIC) of propolis for that specific sample. The minimal concentration of propolis that dropped the curve to the baseline due to killing of bacteria and retained it at that level for the rest of the experimental period (8 to 10 h) was considered as the minimal bactericidal concentration (MBC). In addition to the immediate drop of the curve to the baseline due to treatment of the culture with MBC, few cases were observed where the curve dropped first by 80 to 90% of the initial level and started to decline gradually and reached the baseline in a short while, after 1 to 2 h. Such propolis concentrations were also

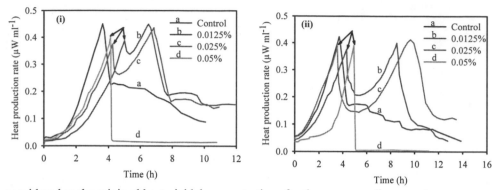

considered as the minimal bactericidal concentrations for the corresponding samples.

Fig. 7.6 Effect of different concentrations of EEP from (i) Poland (P3) and (ii) South Africa (SA8) on the structure of the *p-t* curve of the bacterium *Bacillus megaterium*. Arrows indicate the time of treatment.

In cases where the treatment of a culture with a certain concentration of propolis accelerated the microbial growth and increased the slope of ascend of the curve after treatment significantly (Paired sample t-test, $\alpha = 0.05$), compared to that before treatment, it was not categorized as MIC. Such a concentration was considered as that which encourages microbial growth through a hormesis effect. If a treatment with a weak concentration of propolis did not result in a drop in the level of the *p-t* curve and if the slope of ascend of the curve after treatment was not significantly different from that before treatment, the corresponding concentration was considered to have no effect. The MIC and MBC values for the different propolis samples were determined according to these criteria and are summarized in Fig. 7.7.

The MIC values of most ethanol-extracted propolis samples lay at about 0.005% w/v with the exception of 5 of the 6 South African samples and the samples from Russia (Rus1) and Germany (G1) that showed a MIC value of 0.0125% w/v. The MBC value for most propolis samples was 0.05 % w/v with few exceptions. The sample from Kazakhstan killed bacteria and dropped the curve to the baseline at a concentration 4x weaker than most other samples, even though it had the same MIC value as others. The weakest propolis samples regarding the MBC value were the sample from Russia and 2 from South Africa. These 3 samples already showed relatively higher MIC values.

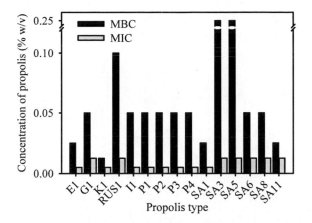

Fig. 7.7 Minimal inhibitory concentrations (MIC) and minimal bactericidal concentrations (MBC) of different propolis samples against *B. megaterium*, determined calorimetrically.

7.4.5.3.2 Level of the *p-t* curve after treatment with different concentrations of propolis

The treatment of an exponentially growing bacterial culture with a concentration of propolis $\geq$ MIC resulted in a drop of the *p-t* curve to a lower level. Based on the type and concentration of propolis, the curve then either stayed at that level for a certain period of time, and then ascended to achieve a second peak or gradually dropped to the baseline, due to bacterial death (Fig. 7. 6 i and ii). By measuring the vertical distance between the point on the *p-t* curve at which treatment was done and the lowest point achieved on the curve due to the treatment of the culture, the dose-response relations for the different concentrations of propolis from various geographical origins were summarized in Fig. 7.8. The dose response curve for each propolis sample is typical, increasing the level of drop in the heat production rate due to treatment with growing concentration of propolis. As observed for all propolis samples, if the level of the curve was reduced by 80 to 100 %, it did not recover, and no second peak was observed. However, except for the sample from Kazakhstan (K1) that initially dropped the curve to 65% of the original value which gradually descended to the baseline, the curves after falling by up to 75% of their original values recovered with time and achieved second peaks. It is not, however, always true that a curve after falling to a lower level may ascend to attain a second peak or gradually drop to the baseline. A different phenomenon was observed with *E. coli* whereby after the curve dropped to 44% of the value at treatment it remained at that level for the rest of the experimental period and started to drop after several hours (Fig. 7.4 i). The length of time a *p-t* curve needed to revive and come back to the same level as before treatment and subsequently achieve a second peak was positively correlated with the level of the drop of the *p-t* curve (Fig. 7.9), which is in turn positively correlated with the concentration of propolis for each propolis sample (Fig. 7.8).

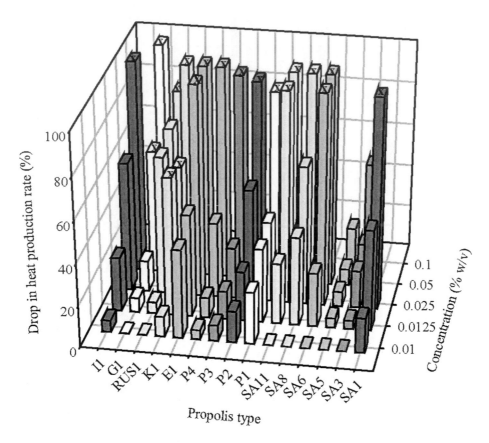

Fig. 7.8 Effect of EEP treatment on the heat production rate of *B. megaterium* demonstrated by the percentage drop in the level of the *p-t* curve. Control treatments with 60% ethanol and distilled water showed no effect. The "x"s at the top of the bars indicate bacterial death and drop of the curves to the baseline suddenly or gradually.

The relations between the level of drop of the *p-t* curve and the time needed for the curve to recover to the same level as before vary from sample to sample significantly as exemplified in Fig. 7.9. Even though significant differences between the various samples existed, all showed the same trend of ascending recovery time with the increase in the level of drop of the *p-t* curve. For this reason, the data for all propolis samples were pooled together and a correlation was made between the pooled drop in the *p-t* curve and the pooled recovery time (Fig. 7.9), and they were positively correlated with a coefficient of determination r^2 = 0.63.

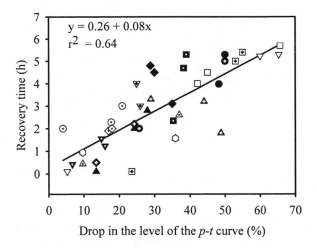

Fig. 7.9 Relation between pooled percentage drop in the level of the *p-t* curve of *B. megaterium* after treatment with 0.0125, 0.025, and 0.05% w/v EEP from different geographic origins, and the time needed for the curve to recover to the same level as before treatment. Pooled results for the 16 propolis samples investigated.

7.4.5.3.3 Slope of curve ascend after treatment

The treatment of an exponentially growing bacterial culture with propolis causes a drop of the *p-t* curve to a lower level, and the curve may ascend again to achieve an after treatment peak. The slope of ascend after treatment could be an important criterion to compare the antimicrobial activities of different concentrations of a sample or a given concentration of various samples, since it could be a reflection of the mechanism of action of the antimicrobial agents.

The treatment with propolis affected the slope of ascend of the curve of a surviving culture (Table 7.5). Actually the length of time needed for the curve to recover to the same level as before is determined by the length of the second lag phase incorporated into the curve due to treatment and the slope of ascend of the reviving curve. The steeper the slope of ascend and the smaller the second lag phase, the shorter is the curve rebound time.

In order to use the slope of ascend as a criterion to compare the effect of different samples and concentrations, the slope of ascend of the exponential phase of growth after treatment was divided through the slope of ascend of the curve before treatment at the exponential phase and multiplied with 100. For almost all propolis samples it was observed that the treatment with lower concentration influenced the slope of ascend less than the treatment with a higher concentration. It was also observed that the treatment with very weak concentrations of some propolis types promoted the growth of *P. syringae* and the curves ascended at significantly higher rates after treatment than before. Typical examples for this phenomenon were the treatments with 0.01% K1, 0.0125 P1, 0.0125 P2, 0.0125 P4 (Table 7.5). The phenomenon of hormesis for *P. syringae* was also observed in the Petridish bioassay experiments whereby dense bacterial growth was encouraged around wells containing concentrations lower than MIC or slightly higher than MIC values. In the latter case the dense growth zone was preceded by inhibition zone around the well (Fig. 7.1, and see appendix Table A1).

Table 7.5 Effect of propolis on the calorimetric power-time curves
Effect of the treatment of an exponentially growing culture of *B. megaterium* on the subsequent features of the calorimetric *p-t* curve displayed by (i) the percentage ascend of the slope of the curve after treatment compared to that before treatment, and (ii) the percentage level of the peak after treatment compared to the level of a control peak. Since higher concentrations of K1 caused lethality, the effect of a 0.01% on the curve is shown here.

	After treatment with 0.0125% w/v EEP		After treatment with 0.025% w/v EEP	
Propolis	% Slope	% Peak	% Slope	Peak
E1	33.8	99.5	declining	none
G1	96.2	100.8	80.4	97.8
K1	138.1 (for 0.01%)	96.8 (for 0.01%)	none	none
Rus1	100.0	98.2	46.3	98.8
I1	95.5	99.7	97.4	100.3
P1	119.6	101.1	65.6	102.5
P2	187.5	103.2	58.5	102.1
P3	72.3	95.5	66.1	95.8
P4	119.4	95.8	38.9	98.4
SA1	70.4	96.1	74.3	98.2
SA3	97.7	104.0	57.3	106.3
SA5	92.6	106.1	106.3	101.5
SA6	37.7	97.5	declining	none
SA8	58.5	98.8	58.9	97.7
SA11	47.8	99.6	declining	none

7.4.5.3.4 Level of the *p-t* peak achieved after treatment

As it was not clear whether the treatments with propolis resulted in different levels of *p-t* peaks, like the different levels of drop and subsequent features of the curves, it was necessary to compare the levels at which peaks of the treated culture *p-t* curves were achieved. As the peak levels of untreated culture of *B. megaterium* were very similar to each other with a mean ± SD value of 0.447 ± 0.004 $\mu W\ ml^{-1}$(n = 5), comparison of individual values with the mean was considered reliable. The treatment peaks lay in the range of 95.5% to 106.3% of the level of the control peaks with a mean of 99.6%, with no significant difference (student's t-test, p > 0.05). No significant differences were observed among the propolis samples or concentrations (2-way ANOVA, $\alpha = 0.05$) (Table 7.5).

7.4.5.4 Comparison of the antimicrobial activities of different propolis extracts

Three different extracts, ethanol-extracted propolis (EEP), water-extracted propolis (WEP) and volatile components of propolis (PV) of the sample G1 were compared using several calorimetric curve parameters to observe if there was any difference in the kinetics of action against *B. megaterium*. The MIC and MBC values for these extracts were, respectively, 0.0125 and 0.05 for EEP, 0.05 and 0.25% w/v for PV, and 0.1 and 0.5 for WEP. EEP required less concentrated solutions to show an inhibitory action and total microbial death than PV followed by WEP. Apart from these differences, no basic difference in the pattern of kinetics of antimicrobial action could be found. The dose-response curves of concentration versus drop in the level of the *p-t* curves due to treatment with the different extracts showed the same pattern but at different concentrations (Fig. 7.10 a). Not only did they show resemblance in the level of drop of the curves but also with the slope of ascend of the reviving *p-t* curves, and the level at which the peaks were achieved after treatment (2-way ANOVA, $p > 0.05$, n = 3) (Fig 7.10 b).

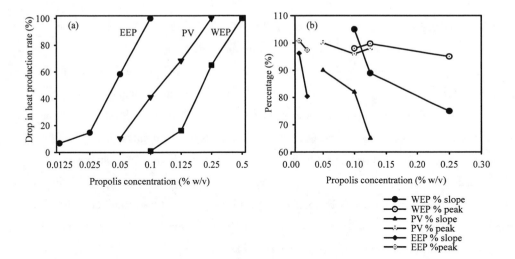

Fig. 7.10 The effect of treatment of an exponentially growing culture of *B. megaterium* with different concentrations of EEP, of PV, and of WEP on (a) the percentage drop in heat production rate after treatment compared to that before treatment, (b) the percentage ascend of slope of the curve after treatment compared to that before treatment and the percentage level of peak after treatment compared to a control peak.

7.5 Discussion

7.5.1 Physico/electro-chemical properties of propolis and yield of extraction

pH values: Unpredictable differences in pH values of the extracts of the different propolis samples exist with no relation to their geographic origin, the species or subspecies of bees that collected them, the yield of extraction, the quality of propolis, and the method of extraction. This electrochemical property of solutions is affected by the quantity and/or quality of acidic/basic substances in the solution. It is rather a reflection of the proportion of positive and negative charges that are found in a solution, and thus samples completely different in their chemical composition may have similar pH values based on the proportion of charges they possess. For this reason the changing pH values between samples even from the same apiary should not be surprising since such samples have already shown difference in the quality and/or quantity of certain chemical groups they possess. Neunaber (1995) demonstrated, using multiple samples from several countries and geographic locations, that the concentration of phenyl-substituted carboxylic acids such as benzoic acid, caffeic acid, ferulic acid, quercetin, cinnamic acid, pinocembrin, pinobanksin, galagin, etc., from the same beehive varies highly based on the year and/or the season of collection. In addition, she also mentioned that the differences between samples from the same apiary may be higher than that between samples from completely different geographic locations. Greenaway et al. (1990) observed a very high quantitative difference (eg. 2 % versus 20% caffeic acid and its esters) in the composition of propolis samples collected 400 meters apart from each other in Oxford. The researchers attributed this difference to the different clones of *Populus sp.* visited by the bees, each clone of the plant being characterized by its own specific chemical composition.

Electrical conductivity: The heterogeneous nature of electrical conductivity of similar extracts of various samples could be explained based on the quantitative and/or qualitative difference in the content of mobile charge carriers (free electrons and/or free ions) in the samples. The larger the total concentration of mobile charge carriers, the higher is the electrical conductivity. A clear difference was observed between the different extracts the strongest value for WEP, and the weakest for PV. This could be due to the fact that water extracts only the charged/polar components of propolis and hence has a higher concentration of free ions (electrolytes) compared to the other two, which possess a bulk of non-polar components with little amount of polar components. The electrolytic potential of polar components in EEP and PV is weakened due the dilution factor by the non-polar groups. Results similar to the findings in this chapter were published by Lindenfelser (1967), and König and Dustman (1988). Due to the physico-chemical heterogeneity of propolis samples from the same location and apiary, the use of electrical conductivity as standardization criteria was put under question by König and Dustmann (1988) and is confirmed here too.

Types of extracts: The extraction of propolis with ethanol helps to procure, among others, flavonoids and phenyl substituted carboxylic acids (Neunaber 1995), of which the flavonoids make up about 95% of the total (König and Dustmann 1988); but wax and fat remain unextracted (Woisky and Salatino 1998). Phenolics (flavonoids and phenyl substituted carboxylic acids) are ubiquitous plant secondary metabolites and hence are found in all propolis samples (König and Dustmann 1988, Greenaway ct al. 1990, Burdock 1998). They make up more than 50% of the total weight of raw propolis (Bankova et al. 1994).

Ethanol (70%) extracts those components soluble in alcohol and water. In addition to that it may also extract the volatile components of propolis. Water extracts only the polar components of propolis such as phenyl substituted carboxylic acids and their esters (e.g. cinnamic acid and caffeic acid that make up to 44% of the total water soluble components of propolis, Neunaber 1995). Other polar components that can be extracted with water include amino acids, and sugars (Marcucci 1995, Nagai et al. 2003). Intuitively it can be deduced that the water soluble components of propolis make up a small fraction of the total extractable components of propolis, as demonstrated in the present investigations too. The lower proportion of the water extractable components was also demonstrated by other researchers ranging from 2.6 to 6.3 w/w (Spiridonov et al. 1992, Neunaber 1995), 2 to 4% for Chinese propolis and 6 to 14% for Brazilian propolis (Miyataka et al. 1997).

The volatile components of propolis (essential oils) consist mainly of terpens and terpenoids (mono-, sesqui-, di-, tri-). Other constituents of volatile oils include alcohols, mainly aromatic alcohols, phenols, aldehydes, ketones, acids (from acetic to stearic acid), esters, a series of alkanes, alkylated benzenes and naphthalene (Bankova et al. 1994). The variation in the volatile oil composition of propolis from different locations is higher than in their phenolic components (Bankova et al. 1994). Since the extraction with water involves heating of the setup, the water soluble components of propolis may also possess the propolis volatiles, but the extraction of PV extracts only the volatile oil components of propolis. For this reason the yield of extraction of propolis is in the order of PV < WEP < EEP. The volatile oil components of propolis were shown to make up 0.2 to 1.28% (Bonvehi and Coll 2000), 0.28 to 0.49% (Woisky and Salatino 1998), 0.27 to 0.60% (Bankova et al. 1994), 0.1 to 0.6% w/w (Kujumjiev et al. 1999) for propolis from different geographic samples. These results are similar to the yield of volatile constituents of propolis displayed in Table 7.1.

The difference in the yield of extraction between different propolis samples is to be expected as it could be affected by the resin content of the material the bees collect from the plant and the amount of wax they mix with it. The resin and wax content of propolis could differ based on geographic location, vegetation composition, bee species and availability of plants (Meyer 1956, Johnson et al. 1994). In seasons and locations where propolis source plants are scarce, the colony suffers propolis shortage and bees were observed collecting "propolis substituents" like asphalt, paint, mineral oils (König 1985). These propolis substituents are then mixed with the available propolis resin and used in the beehive (König 1985). In addition to the non-propolis materials collected from the environment and mixed with propolis, the bees can also control and vary the proportion of wax they add to the plant resin based on the availability of plant resin and the purpose for which propolis is to be used (Burdock 1998). For example propolis used to repair honeycombs is often supplemented with large quantities of wax to give it a firmer composition, while propolis applied in a thin coat to the surface of a comb usually contains little or no wax (Meyer 1956). Thus the lower yield of some propolis samples that looked pure up on visual inspection could be contributed to the higher wax content which can not be extracted in 70% ethanol.

The lower yield of "impure propolis" samples is due to, among other factors, the higher proportion of foreign materials that the bees collect and use as propolis substituents, and also substances incorporated to propolis by the beekeeper during the collection of propolis. The quality of propolis is dependent on how propolis is procured from the frames, cover boards, collection matt, and the inner wall of the beehive (Bogdanov 1999). The stingless bee, *Tetragonisca angustula* Illger is normally known to collect propolis and mix it with wax and soil to form geopropolis (Bankova et al. 2000). Thus soil is the normal component of geopropolis from stingless bees imparting lower yield to this type of propolis. A comparative investigation on the chemical make-up of propolis from stingless bees and honeybees from the same geographical location showed that the two propolis samples have similar yield and also resemble in chemical composition with slight differences that could be traced to the preferred trees visited by the two bee species (Pereira et al. 2003).

7.5.2 Petridish bioassay of different propolis samples

The results of antimicrobial tests are unambiguous proofs, that in spite of the great difference in the chemical composition of propolis of different geographical origins and collected by various bee races, all of them exhibit significant antibacterial and antifungal effects. But the strength of antimicrobial activity could differ based on the nature of the specific substances in each sample. The differences in the strength of antimicrobial action between the various propolis samples may not be surprising since the composition and proportion of the bioactive components in the different samples differ from each other quantitatively and/or qualitatively. Kedzia et al. (1990) proposed that the antimicrobial action of propolis is complicated and could be due to the synergism between flavonoids, hydroxyl acids, and sesquiterpenes. It was experimentally demonstrated that not even a single component isolated from propolis showed an activity higher than the total extract (Metzner et al. 1977, Kujumgicv et al. 1993, Bonvehi et al. 1994). The synergistic effect between the different components of propolis was already reported by Scheller et al. (1977), and latter confirmed by Krol et al. (1993). It is thus obvious that, in different samples, different substance combinations are essential for the biological activity of propolis, rather than only one, and hence samples of completely different geographic origins may have comparable antimicrobial activities (Kujumgiev et al. 1999).

Though no chemical analysis was performed to identify the bioactive components in the samples used in the present investigations, it is possible, based on literature data of propolis from different geographic origins, to state that the samples do differ in their chemical composition. The degree of variation depends on the type of plants visited by the bees, which is in turn determined by the geographical location (for example Ghisalberti 1979, Marcucci 1995, Miyataka et al. 1997, Burdock 1998, Kujumgiev et al. 1999, Bankova et al. 2000, Bankova et al. 2002). Though the specific chemical make-up of the plant resins collected by the bees differs based on the plant species visited, the general framework of bioactivity is more or less the same. Regardless of the geographic location, where a plant grows, the purposes for which it secretes resin resemble at least partially. Resin is mainly secreted by plants in order to seal wound, to stop sap loss and protect wounds from infection by microbes, to protect against infection of pollen (it is coated with resin), to stop germination of seeds and sprouting of bud while frost (Ogren 1990). Though the specific chemicals that are responsible for these actions could differ, the essence of action remains the same, leading to the similar biological activity of different samples.

7.5.3. Comparison of the sensitivity of different microbes to propolis

All of the Gram positive bacteria tested were highly sensitive already to lower concentrations of propolis, but the Gram negative bacterium *E. coli* displayed a reduced sensitivity to most of the samples and was insensitive to 2 of the 16 samples tested. The lower sensitivity of *E. coli* is in agreement with findings by several researchers where this bacterium showed either very low sensitivity or total insensitivity against propolis (Digrak et al. 1995, Marcucci 1995, Kujumjiev et al. 1999, Bonvehi and Coll 2000, Sforcin et al. 2000). It can not, however, be generalized that Gram negative bacteria are insensitive to propolis since the other Gram negative, *P. syringae*, even though it showed relatively higher MIC values, had a sensitivity similar to most Gram positive bacteria at higher concentrations of propolis. But it can be ascertained that the Gram negative bacteria are less sensitive than the Gram positive ones, at least at lower concentrations of propolis, which is in agreement with literature data (Grange and Davy 1990, Dobrowolski et al. 1991, Sforcin et al. 2000). The most plausible explanation for the less sensitivity of Gram negative bacteria is their outer membrane that inhibits and/or retards the penetration of propolis at lower concentrations. After a certain threshold concentration, however, the impermeability of this membrane is disturbed resulting in the movement of bioactive components of propolis into the cell interior, resulting in cell inhibition or death. Another possible reason why the Gram negative bacteria are more resistant to propolis might be the possession of multidrug resistance pumps (MDRs), which extrude amphipathic toxins across the outer membrane (Tegos et al. 2002). The presence of MDRs in *E. coli* and their role in the insensitivity of the bacterium to antimicrobials was clearly elucidated (Lomovskaya and Lewis 1992, Nikaido 1999, Zgurskaya and Nikaido 1999, Lewis and Lomovskaya 2001). The MDRs could be very effective at lower concentration of propolis and extrude those molecules that crossed the outer membrane barrier. This pumping potential of the MDRs can be saturated at a certain threshold concentration and the rate of penetration and accumulation of antimicrobials in the cell interior increases, resulting in antimicrobial effect at higher concentrations.

Fungi are generally less sensitive than bacteria in terms of the MIC values and/or diameter of inhibition zones at higher concentrations, except for the yeast that showed higher diameters like that of the bacteria. Considering the MIC values, the yeast had a sensitivity in between the highly sensitive bacteria and the less sensitive mould.

The saturation in the dose response curves for the filamentous fungi, with little change in the diameter of the inhibition zone as the concentration increases could be due to the weak solubility/insolubility of propolis in the agar layer. The weak diffusion potential of propolis in agar layer and hence unexpectedly small diameters of inhibition zones were observed by other researchers too, when using higher concentrations of propolis against fungi (Metzner et al. 1979, Sawaya et al. 2002). The strongly non-polar components of propolis precipitate on the inner wall of the agar hole, or after diffusing a very short distance, blocking the diffusion path and retarding migration of potentially diffusible substances. So, for the diffusion of propolis at higher concentrations, the general principle of diffusion, that rate of diffusion increases with increasing concentration may not work. After the end of the experiments, all Petridishes containing higher concentrations of propolis such as 4%, 8% and 10% displayed propolis precipitations at the inner wall and on the bottom of the hole.

The presence of certain bacteriostatic/bactericidal chemicals in the growth medium at a concentration lower than a critical inhibitory level could enhance the growth of an organism that otherwise would have been inhibited/killed by higher concentrations, a phenomenon known as hormesis (Edward et al. 1998). The phenomenon of hormesis was displayed by *P. syringae* at concentrations of propolis lower than MIC for most propolis samples. This situation alerts that if propolis is to be used in treatment of infections it has to be used at concentrations far above the MIC values, in order to minimize the risk of encouraging bacterial growth at or in the immediate surrounding of the site of application.

7.5.4 Comparison of the antimicrobial activities of different propolis extracts

EEP showed the highest antimicrobial activity compared to the other two extracts, even though the differences between EEP and PV were not significant for most samples. The reason why EEP is superior to WEP, and in some case to PV is that extraction of EEP extracts all water and ethanol extractable and biologically active components that are also present in the other two extracts. In addition, EEP contains several bioactive components that are not found in WEP and PV. Since it is already pointed out that the antimicrobial activity of propolis could involve a synergistic interaction between its components (Scheller et al. 1977, Kedzia et al. 1990, Krol et al. 1993, Kujumgiev et al. 1999) the possession of higher activity by EEP would not be surprising. Regardless of the possession of few types of biologically active compounds (Kujumgiev et al. 1999) the volatile components, astonishingly showed activities comparable to EEP. This could be due to the fact that even if PV contains fewer types of compounds compared to EEP, it may contain a sufficiently high amount of bioactive compounds to result in an effect comparable to the former. It is known that the PVs possess good to moderate antibacterial activities and they are responsible for the lower incidence of microbial aeroflora within the apiary (Ghisalberti 1979, Petri et al. 1986). Since the rest of propolis is insoluble in the beehive interior and hence "inactive", the volatile components of propolis are essential to reduce infections in the bee hive. The yield of extraction and consequently the antimicrobial action of the volatile components depend on the age and storage situation of propolis (Woisky and Salatino 1998). If it is not stored in a tight container, components are lost with time and the sample loses its potential. This may be the reason, apart from the very nature of chemical make-up of the various propolis samples, for the different yield and antimicrobial activities of the PVs from various samples investigated in this study.

7.5.5 Calorimetric experiments

The results of the Petridish bioassay experiments were affected by the insolubility of propolis in the agar layer, especially while using highly concentrated propolis against the relatively insensitive fungi. The insolubility problem was not serious at lower propolis concentrations tested against bacteria since the highly diluted hydro-insoluble components could diffuse through the agar layer with the excess solvent in which they are dissolved (ethanol). The problem of insolubility could still be there at lower concentrations and affect the diffusion potential, though not clearly observable. The calorimetric results, however, were not and can not be affected by this problem since it is done in a nutrient broth. Due to vigorous stirring of the culture, the water insoluble components of propolis remain suspended in the medium and express their antibacterial activity.

One of the limitations confronted with calorimetric experiments was that the calorimetric spiral, where heat production is detected, and the fermenter where cultivation of cells is done, are separated by an unavoidable geometric distance. This distance is necessary since a minimum tube length is required for the precise thermal equilibration of the culture before it arrives in the calorimetric spiral. The culture in the fermenter is aerated continuously, and as a result, the bacterial growth is not limited by oxygen. Growth of the bacterial culture thus continues until carbon and energy supplies in the fermenter are exhausted, marked by a declining phase of count of CFU. The situation in the tubing system and hence the calorimetric setup is, however, different. As a culture sample leaves the fermenter and enters the tubing system it possesses a certain concentration of oxygen, which is continuously consumed and reduced by the metabolizing bacteria. At a low cell density, bacterial growth and metabolism continue, since there is sufficient oxygen for the bacteria to utilize, even though the concentration of oxygen decreases continuously. But, this situation changes at higher cell densities. The available and 'in the flow line non-replenishable' oxygen is consumed nearly totally before the culture arrives at the calorimetric spiral, and hence bacterial metabolism and heat production rate drop drastically in the tubing system. In case of the facultative anaerobe, *E. coli*, a different situation was observed. Upon exhaustion of oxygen in the flow line the metabolic rate started to drop until it reverted and ascended again because of fermentation. Fermentation was initiated after the online oxygen concentration dropped to about 30 µmol l^{-1}. The anaerobic curve ascended further and achieved a plateau level at about 0.5 µW ml^{-1} and remained at this level for the rest of the experimental period until it started dropping due to complete exhaustion of nutrients (carbon, nitrogen and energy sources). The constant level of heat production rate indicates that there is no significant bacterial growth taking place, and that the heat output is mainly due to maintenance metabolism, not of growth. This fact is also demonstrated by the nearly constant level of CFU in the anaerobic phase.

Due to depletion of oxygen in the flow line the declining calorimetric curve may not be a good indicator of the metabolic activity in the fermenter itself after a growth peak is achieved in the calorimeter (Hölzel et al. 1994, Garedew and Lamprecht 1997, Garedew et al. 2003). The calorimetric death phase may actually be the late exponential as well as the stationary phase of bacterial growth in the fermenter. Therefore, killing some bacteria in the fermenter and the flow line by applying a moderate concentration of propolis may open an opportunity for the survivors to utilize the excess oxygen and nutrients available, thrive and increase the heat production rate showing the true picture of the phenomena that take place in the fermenter. The treatment at this phase could also help to find out whether the drop in the heat production rate was caused only by the shortage of oxygen in the flow line or due to exhaustion of nutrients in the fermenter and flow system as a whole. If the drop of the heat production rate was caused by exhaustion of nutrients, the treatment with propolis should not result in a boost of the heat production rate. But the treatment with a very strong concentration of propolis may kill the bacterial culture in the fermenter and no heat production could be expected. For this reason this treatment was done only with moderate concentrations of propolis.

Therefore, due to the above mentioned problems associated with length of the tubing system and subsequent oxygen depletion, calorimetric recording of the heat production rate is a true picture of events taking place in the fermenter only at lower cell densities. If calorimetric data are to be used at higher cell densities, the results have to be compared with other data, such as the oxygen concentration in the flow line and the number of CFU, and results have to be interpreted with caution.

It has been well documented that the sensitivity of bacteria to antimicrobials is highly affected, among other factors, by the growth phase at which treatment is done (Hogan and Kolter 2002). Slow growth rates of bacteria resulting from nutrient limitations are accompanied by an increased resistance to multiple antimicrobial compounds because of low metabolic activities and decreased cell permeability (Brown and Williams 1985).

Furthermore, slow growth is also correlated with the increased expression of multi-drug efflux genes (Rand et al. 2002). A recent research finding showed that *E. coli* stationary phase cultures produce a diffusible factor that induces resistance to antibiotics in growing *E. coli* cells but the nature of the signal and the response of the cells is not yet clear (Heal and Parsons 2002). It has also been proposed that, in stationary phase cultures, there are subpopulations of antibiotic resistant 'persister cells' that lead to the increased resistance to antibiotics (Spoering and Lewis 2001). Persister cells are not mutants and do not produce antibiotic resistant offspring in the absence of antibiotics. It is not yet known how persister cells are formed or what factors contribute to their antibiotic resistance (Hogan and Kolter 2002). Due to these facts, treatments of the bacterial cultures were done at the exponential growth phase, before the calorimetric *p-t* peaks were reached.

The treatment of bacteria with propolis in the calorimetric exponential phase resulted in a drop of the heat production rate to a lower level, depending on the concentration of propolis. If the concentration is weak, it does not kill all bacteria, and hence the survivors do continue to metabolise, maintaining the heat production rate at a certain level above the baseline. This level is kept for some time directly depending on the concentration of propolis, after which the curve revives and ascends again. The most plausible explanation for this behaviour could be that, after treatment a certain proportion of the cells is killed, others are inhibited, and some others remain unaffected. The metabolic heat production rate in this case could originate from the inhibited cells, since they could perform maintenance metabolism, and those of the unaffected and normally metabolizing and growing cells. It may be possible that the number of normally metabolizing cells is very low and the change in heat production rate is below the detection limit of the calorimeter. But with the course of time, the number increases and the change in their metabolic rate could be detected on the *p-t* curve. The survivors then metabolise and grow in the flow calorimeter displaying the *p-t* curve typical for the bacterial strain. Another possible explanation might be that some bacteria could be inhibited and perform their maintenance metabolism for a certain period of time. The inhibition may then come to end, and the bacteria start metabolizing and growing again displaying features typical of a normally growing bacterial culture. The second hypothesis agrees with the proposal by Maillar (2002) that bacteriostatic effects achieved by lower concentration of biocides might correspond to a reversible activity on the cytoplasmic membrane and/or impairment of enzyme activity.

The treatment of bacteria with EEP dropped the heat production rate and raised the online and fermenter oxygen concentration immediately, but the drop in the number of CFU was gradual. A plausible explanation for this could be that immediately after treatment, the metabolic rate and oxygen consumption drop drastically even though the organisms were not dead, but weakened. Removal of a sample and culturing it on propolis free medium releases the bacteria from the antimicrobial agent and they start growing. But with the increase of incubation time with propolis in the fermenter the number of dying bacteria is increasing and hence the CFU curve declines. The results of the present investigation are in agreement with that of Sforcin et al. (2000), whereby incubation of a sensitive *S. aureus* strain and a relatively less sensitive *E. coli* one in broth containing EEP, and subsequent removal and plating of the samples displayed different patterns of incubation time versus survival curve, with the sensitive bacteria dying in short incubation time.

Since the *p-t* peaks after treatment were achieved at the same level of heat production rate and oxygen concentration in the flow line, as for the control culture, it can be ascertained that the treatment with propolis did not affect the level of oxygen sensitivity of the survivor bacteria. The lower slope of ascend of the *p-t* curve after the treatment with a higher concentration of propolis could be an indication of partial inhibition of cell division of the survivors thus growing and metabolizing at slower rates. The effect of propolis on cell division has already been confirmed (Takaisi-Kikuni and Schilcher 1994).

The calorimetric method displayed MIC values lower than the Petridish bioassay method. The main reason for this could be that the calorimetric method is an online recording of activity and hence even the slightest and short lasting effects could be detected. But in case of the Petridish bioassay method results are cumulative effects of incubation for 24 to 48 h. In addition to that, several other problems encountered by the Petridish bioassay method do not affect the calorimetric results. Results from the Petridish assay method are influenced by inoculum size, agar layer consistency, incubation temperature, polarity and diffusion potential of the test substance. At a first glance, it seems that the inoculum size affects the calorimetric results too, but since treatment can be done at a level of the exponential phase of the *p-t* curve, which corresponds to an exactly predetermined density of CFU, this plays no role. Nevertheless the size of the inoculum affects the length of the lag phase, which has nothing to do with the treatment at the exponential growth phase.

The phenomena of change in the structure of the curves after treatment at the calorimetric death phase could be attributed to the shortage of oxygen in the flow line that resulted in a drop of the *p-t* curve to the minimal point after the peak. The treatments with propolis inhibit/kill some cells in the fermenter and flow line, increasing the availability of oxygen for the survivors, which results in an increased metabolism and ascend of the curve.

7.5.6 Possible mechanisms of propolis action

Apart from the proposal and speculation by several researchers as to the mechanism of action of propolis, no much work has been done, and literature data are scarce. It was pointed out that the mechanism of action of propolis is complex and no analogy can be made to any classic antibiotics (Kedzia et al. 1990).

The slow growth rate of bacteria and the lower slope of ascend of the p-t curve after treatment with a moderately concentrated propolis solution is an indication that the rate of growth/cell division and metabolic activity of the survivors were partially inhibited or reduced. Electron microscopic investigation of propolis treated bacterial cells by Takaisi-Kikuni and Schilcher (1994) demonstrated that bacterial cell division was inhibited and they argued that the mechanism of action could be like that of nalidixic acid which is known to inhibit DNA replication and subsequent cell division. The researchers also argued that propolis could inhibit DNA dependent RNA polymerase. It was also displayed that treatment with propolis interferes with protein synthesis, proved by measuring the concentration of extracellular proteins of control and treated cells (Simuth et al.1986, Takaisi-Kikuni and Schilcher 1994)

The relatively lower sensitivity of Gram negative bacteria compared to the more sensitive Gram positive ones, at least at lower concentrations, could be traced to an activity that involves the Gram positive bacterial cell wall, inhibition of cell wall synthesis and hence distortion of its integrity. Electron microscopic pictures displayed that propolis treated cells possessed defective cell walls and failed to separate after cell division and formed a pseudo-multicellular structure (Takaisi-Kikuni and Schilcher 1994). In an experiment with a known antibiotic, Nishino et al. (1987) demonstrated that the formation of a pseudo-multicellular structure after treatment could be due to the blockage of the so called splitting system of the cross wall. Treatment of *S. aureus* with a combination of benzylpenicillin and propolis showed a synergistic effect witnessing that the effect of propolis could involve inhibition of cell wall synthesis (Schub et al. 1981).

Electron microscopic pictures of treated cells displayed that the cytoplasm of the cells was disorganized and it was emptied after a prolonged incubation time (Takaisi-Kikuni and Schilcher 1994).

Antifungal activities of propolis are supposed to be like that of amphothericin B, which forms complexes with sterols (ergestrol) of the fungal membrane (Cizmarik and Trupl 1975, Pepeljnjack et al. 1982).

It can be concluded that the present results are unambiguous proofs that in spite of the enormous differences in the chemical composition of propolis from different geographic locations, all samples exhibited significant antibacterial and antifungal activities. Hitherto investigations dealing with chemical composition and biological action of propolis did not point out one individual substance or a particular substance class which could be responsible for this action. Obviously a synergistic interaction between different substance combinations is essential in different samples for the biological activity of bee glue. It seems that the chemical nature of propolis is beneficial not only to bees but have general pharmacological values as an antimicrobial natural product.

8.0 General discussion / Allgemeine Diskussion

8.1 General discussion

The large number of bees living in a crowded space can have disastrous consequences for the colony, due to the very fast transmission of parasites and pathogens among individuals in the beehive. The death threat of honeybee colonies attacked by parasites, pests, or pathogens is so high that affected colonies have to be treated in someway to prevent them from perish. The most urgent problem of apiculture based on the western honeybee *Apis mellifera* L. today is the threat of extermination by the parasitic mite *Varroa destructor* (Anderson and Trueman). Parasitic mites and mite-related diseases have caused the death of most wild honeybees and left the commercial colonies at high risk. In addition to parasitic mites, the honeybee is also a victim of attack by several types of viruses, bacteria, fungi, protozoans, and insect pests such as the greater wax moth *Galleria mellonella* L., and the small hive beetle *Aethina tumida* Murray among several others. The effect of the other bee diseases is nowadays overshadowed by varroosis, except the newly emerged, and frightening small hive beetle *Aethina tumida,* which is already ravaging the beekeeping industry in the United States and Australia; but not yet detected in Europe. Varroosis, however, still holds the record of current research activities in the field of bee health.

Even though infestation of a colony with *Varroa destructor* is known to elicit colony death, the exact cause of colony death remains controversial. Some researchers, undermining the amount of hemolymph the mites rob from bee brood, reached the conclusion that the causes of colony death after infestation by Varroa mites are the viruses vectored by them, but not the mites themselves. Apart from guesses and speculations, no data exist about the amount of hemolymph and energy mites rob from the brood during the entire capped developmental period. For the first time, the nutritional and energy demand of mites have been demonstrated in this thesis, through starvation and calorimetric experiments. It has been found that the mites suck up to 28% of the non-replenishable reserve food of the capped brood that otherwise would have been consumed by the pupa during metamorphosis, and would have contributed to proper development. The shortage of substances essential for ontogenesis leads to malformation and underdevelopment. This results in the emergence of underweighted adult bees incapable of performing normal life activities. Therefore, the very nature of loss of high amount of reserve food by the capped brood, due to robbing by the mites, may lead to the emergence of bees with the typical mite syndromes. It can not, however, be excluded that factors other than the Varroa mites themselves, such as the viruses and bacteria vectored by them, could contribute to the malformation of bees infested with

Varroa mites during brood development; the mites, however, still play remarkable roles to cause the problem, or add up to it.

Regardless of whether the mites themselves, or the viruses and bacteria that they vector cause the problem of colony death, mites have to be controlled. Research is being carried out in different laboratories to select for honeybees resistant to mites, and to enable them to live together with the mites by suppressing the mite's population size. In addition to the effort to select resistant bee races, researchers are also focusing on the biology of the Varroa mites, trying to understand how they locate bees in the first place. If one can determine the host location mechanism, and discover the physical and chemical cues the mites use, it may be possible to manipulate those cues for a control mechanism that will protect the bees. The process of selection of honeybees resistant to mites, and understanding the biology of the mites is taking undesirably longer time. Therefore, the colony has to be treated in the meantime to stop its death.

Different types of biotechnical and chemical mite control methods are available at present. The biotechnical methods are laborious, time consuming, and also not very efficient. Effective mite control may be achieved by the use of certain acaricides, which are unfortunately associated with several drawbacks. Some acaricides are toxic to the bees and the beekeeper above a certain threshold concentration, and they also cause residue problems in bee products such as honey, wax, and propolis. The other major problem with the use of acaricides is that mite populations resistant to acaricides have already emerged, thus, acaricides are losing their efficiency. As living organisms, mite populations will eventually adapt to whatever chemical control mechanism one develops, and thus it is an ongoing struggle that may never be completely won. Thus with one solution to varroosis at hand, the search for other, better, and effective solutions could save the honeybee colony from its demise. Presently, the search for antivarroa agents is mainly concentrating on in the realm of organic acids and plant secondary metabolites. A number of natural products, including essential oils from herbs and spices, are being examined for their potential in mite control. It is actually desirable to use a natural product, which is a mixture of several compounds, with different modes of action, since the development of resistance against such mixtures takes longer time. One of such natural products, known for its broad spectrum of biocidal actions, is propolis.

Even though it naturally occurs in the beehive, and does not incur the beekeeper extra cost, the potential antivarroa use of propolis is completely forgotten. The beekeepers and bee researchers know about the antibacterial, antifungal, and several other curing effects of propolis, but not its antivarroa actions. It seems that there is a rationale for bee researchers and beekeepers not to think of the potential acaricidal use of propolis. The *in situ* presence of propolis in the beehive, where the Varroa mites wreck havoc, may lead at a first glance to the notion that propolis does not have any effect on mites; otherwise the mite would have been killed by propolis. In this thesis work the antivarroa actions of propolis of different geographic origins were investigated.

Propolis showed Varroa narcotizing and varroacidal effects. The length of narcosis and the mortality rate of mites depended on the solvent of extraction (70% or 40% ethanol, or water), concentration of propolis, and contact time. Propolis extracted in 70% ethanol was found to be highly toxic, resulting in 100% narcosis of mites regardless of the contact time and concentration of propolis. After treatment with weak concentrations of propolis (e.g. 0.5%, 1%, and 2%) most mites recovered from narcosis, and few died. However, the treatments with concentrated propolis solutions (e.g. 5% and 7.5%) led to higher mortality and little recovery rates from narcosis. The treatment with 40% ethanol-extracted propolis was less effective, both in its Varroa narcotizing and varroacidal actions, compared to that of propolis extracted in 70% ethanol. The treatments with water-extracted propolis had negligible effects on mites, considering their Varroa narcotizing and varroacidal effects.

From the trends observed in the antivarroa actions of the different extracts of propolis, it was clear that extraction with strongly concentrated propolis procures the biologically active hydrophobic components of propolis, that have strong antivarroa actions. The water soluble components of propolis are less active and constitute a minor proportion of its chemical make-up. Propolis is, thus, inactive in the beehive interior *in situ* due to the insolubility of its biologically active components; its potential is concealed by its insolubility. Thus, propolis can be used as an antivarroa substance after extracting and dissolving it in ethanol. Spraying honeybees with Varroa mites on their surface, with 10% propolis in 55% ethanol, in a preliminary experiment, showed that the mites were dead at the end of the experiment. The dead mites either remained attached to the bee or fell down, but the bees were neither narcotized nor aggressive. This is an indication that propolis solution can be used in the beehive interior as a varroacide.

Even though the treatment with lower concentrations of propolis did not result in a remarkable mortality of mites, it significantly reduced their heat production rate. This indicates that even the non-lethal doses of propolis could be used in mite control, instead of the strongly concentrated and lethal ones. The use of lower concentrations of propolis is desirable, since higher concentrations, though they showed no observable effects in a preliminary experiment, may affect the bees in some way. It is also wastage of resources to use higher concentrations of propolis, if the desirable effects could be achieved with the lower ones. The effectiveness of treatments at lower concentrations can be increased by exploiting the phenomenon of synergism of propolis treatment with high temperature (e.g. 40 °C).

The antivarroa effects of propolis investigated in this thesis concentrated on its narcotic effect, mite weakening, reduction of the heat production rate, and varroacidal actions. It is, however, possible that propolis affects the chemical cues and orientation of mites, thus disturbing their ability to locate, and enter a ready-to-cap brood. This latter mechanism was displayed by some essential oils.

Investigations on the antivarroa actions of ethanol-extracted propolis from different geographic origins displayed that all propolis samples showed antivarroa actions, with slight differences in the strength of narcotic and varroacidal effects. The slight differences in the strength of activity of propolis samples of different geographic origins can be explained by the quantitative and/or qualitative variation of the propolis chemical make-up. The geographic location, and, therefore, the vegetation composition of the area, affects the chemical make-up of propolis quantitatively and/or qualitatively. Some propolis samples, however, showed similar antivarroa activities regardless of their geographic origins. The possession of comparable antivarroa actions by the different propolis samples is an indication that, regardless of the origin, propolis is collected and accumulated in the beehive for similar purposes.

Not only the mite *V. destructor*, but also a weakened-colony-devastating, and storehouse comb destructive insect pest, the greater wax moth *Galleria mellonella* that was also sensitive to propolis treatment. The treatment of the wax moth larvae with propolis narcotized them, and reduced their heat production rate remarkably. The sensitivity of wax moth larvae to the propolis treatment changed with larval instar, with the early instars displaying higher sensitivity. Higher concentrations of propolis induced mortality of larvae. Even though weak concentrations of propolis did not kill the larvae, they affected metamorphosis of the pupae that were treated at the seventh larval instar, with the strength and nature of the effect being dependent on the sublethal concentration of propolis.

After treatment with propolis concentrations that are sublethal to the seventh instar larva, all treated larvae underwent larval-pupal ecdysis, with highly diminished ecdysal peaks. The nature of the subsequent pupal metamorphosis of such treated larvae was highly affected by the concentration of the sublethal propolis used. After treatment with very weak sublethal concentrations of propolis (e.g. 1% w/v), the pupae were able to complete their metamorphosis. In such cases, metamorphosis lasted shorter than that of the control organisms, indicating that propolis at such concentrations plays the role of insect growth regulators (IGR). The adults that emerged after the accelerated metamorphosis of pupae displayed weakly structured *p-t* curves, which is an indication of their inability to fly properly. The weak flying activities of adults could be explained by the underdevelopment of structures, such as flying muscles. Even though the larvae treated with higher sublethal concentrations of propolis (e.g. 4% w/v) underwent the larval-pupal ecdysis, pupal metamorphosis was aborted. This is a good indication that optimal concentrations of propolis, though not lethal to the larval stage, can be used to control wax moths by interfering with pupal metamorphosis. The effects of weak sublethal concentrations of propolis on the development of *Galleria* are very difficult to judge visually, since no death was observed, and the organisms appeared normal. The calorimetric method, however, helped to monitor the weakened metabolic activities that take place during metamorphosis, and the flying activities of adults.

The screening of insecticidal, insectistatic, and antivarroa actions of propolis by the use of the calorimetric method is highly reliable compared to the traditional and standard methods. The only limitation of the calorimetric method to investigate the insectistatic, insecticidal, and antivarroa actions of propolis is that it was not possible to record the heat production rate of the organism in the first 30 to 45 min after treatment, due to the time needed for the thermal equilibration of the calorimeter. This problem can, however, be solved by coupling the calorimetric method with infrared CO_2 analysers, or oxygen sensors that measure oxygen concentration in air.

The antimicrobial activities of the different extracts of propolis from various geographic locations were compared by using parameters such as the minimal inhibitory concentrations (MIC), minimal bactericidal concentrations (MBC), diameter of the inhibition zone, and several parameters of the *p-t* curves. The most important features of the *p-t* curves used for comparisons were the level of drop of the curve after treatment, the time needed for the dropped curve to revive and to come back to the before-treatment position, and the level of the *p-t* peak achieved after treatment. The ethanol extract of propolis (EEP) was found to be highly effective against all microbes tested. The dose-response curves of concentration versus the effect on the *p-t* curve, by the different extracts, demonstrated that the response patterns were similar to each other, but achieved at lower concentrations of EEP, and very high concentrations of WEP, intermediated by those of the PV. For some propolis samples, there were no remarkable differences in the antimicrobial activities of the EEP and PV. The higher antimicrobial activities of the PV could be an indication that these fractions of propolis play roles in reducing the microbial flora in the air within the beehive. The water extracts of propolis (WEP) were, however, the least active of all extracts, their antimicrobial activities being detected only at higher concentrations. The lower activity of WEP, and its less abundance in the beehive, demonstrated by the low yield of water extracts, explains why this component of propolis does not play considerable roles against the parasites and pathogens of honeybees *in situ*.

The reason why EEP is superior to WEP and in some cases to PV, in its antimicrobial activities, is that the extraction of propolis with ethanol procures all water soluble, ethanol soluble, and the volatile components of propolis; making EEP superior to the other two extracts qualitatively and/or quantitatively. The qualitative and eventually quantitative richness of EEP in its chemical make-up is indicated by its higher yield of extraction, several folds higher than that of PV or WEP.

Since the yield of extraction of WEP is very low, and it has very weak antimicrobial activities, compared to EEP, it is not economical to use WEP in antimicrobial propolis therapy. It can, however, be used at higher concentrations in situations where an alcohol solution of propolis can not be used, such as for religious reasons. The PV have very low yield, compared to EEP, and it also possess activities comparable to or lower than that of EEP. Thus, it is not economically feasible to use PV as antimicrobial agents as far as EEP can be used in place.

The Gram positive bacteria were generally found to be more sensitive to propolis treatments than their Gram negative counter parts and the fungi. Among the two Gram negative bacteria tested, *E. coli* was the most resistant. The other Gram negative bacterium, *P. syringae*, was relatively sensitive to most treatments when considering the diameters of inhibition zones at higher concentrations, but it needed higher MIC, compared to that of the Gram positive bacteria. The lower sensitivity of Gram negative bacteria can be accounted for by the impermeability of their outer membrane to antibacterial agents. The especially low sensitivity of *E. coli* to propolis can be due to the presence of multiple drug resistance pumps, in addition to the impermeability of the outer membrane, which are already confirmed for several strains of this bacterium against different types of drugs.

The three filamentous fungi were less sensitive to propolis treatment than the yeast *Saccharomyces cerevisiae*, which showed higher diameters of inhibition zones, especially at higher concentrations of propolis. The sensitivity of *S. cerevisiae* to concentrated propolis solutions is comparable to that of most Gram positive bacteria, except that the yeast demonstrated relatively higher MIC values.

Investigation of the antifungal actions of propolis using the Petridish bioassay method is highly limited by the weak solubility of propolis in the agar layer. This fact was demonstrated by the saturated dose-response curves of propolis concentration versus diameter of inhibition zone of all filamentous fungi. Though retardation of propolis diffusion through the agar layer also occurs in the bacterial Petridish bioassay experiments, it is not as sever as in the case of fungal cultures. The reason for this could be the higher consistency of fungal growth media that greatly impedes diffusion as compared to the less consistent bacterial growth media.

The problem of propolis diffusion across the agar layer can be solved by using nutrient broth rather than nutrient agar, which is, however, applicable only for bacterial and yeast cultures. Due to the colour of propolis and the formation of emulsions upon addition of propolis to a bacterial culture in a broth, its antimicrobial action is impossible to investigate spectrophotometrically. The use of the calorimetric method, however, solves these problems since it is not affected by the colour of the media. The calorimetric method is more sensitive than the Petridish bioassay method, since it displays antibacterial actions of propolis concentrations considered to be non-inhibitory by the other method. In addition to enabling us to detect the activities of lower concentrations of propolis, the calorimetric method contributes in the elucidation of the mechanisms of the biocidal actions of propolis, since it records the metabolic activities of bacteria "online". The Petridish bioassay method, in contrast, displays cumulative results of incubation for 24 to 48 hours, or more. Based on the calorimetric results it was possible to illustrate that propolis possesses concentration-dependent bactericidal and bacteriostatic actions.

The calorimetric method, however, is also not without drawbacks. The limitations of this method arise due to the tube connection between the fermenter and the calorimetric spiral. Due to the tube connection, the oxygen concentration in the flow line drops drastically at higher cell densities, introducing artefacts in the p-t curve. However, as far as treatment is carried out before the p-t peak is achieved, the curve is a reflection of the events that happen in the fermenter, making the calorimetric method very reliable and enabling us to exploit its higher sensitivity in the investigation of the antimicrobial action of biocides such as propolis which are difficult to investigate using other methods.

8.2 Allgemeine Diskussion

In einem Bienenstock lebt eine hohe Zahl von Bienen auf beengtem Raum zusammen. Das hat die schnelle Übertragung von Parasiten und Pathogenen zur Folge. Viele der infizierten Völker sterben ab. Das dringlichste Problem bei der Zucht und Haltung der westliche Honigbiene, *Apis mellifera* L., stellt deren Befall durch die parasitische Milbe *Varroa destructor* (Anderson und Trueman) dar. Darüber hinaus sind die Honigbienen durch verschiedene Virusstämme, Bakterien, Pilze, Einzeller und parasitische Insekten, z.B. die Große Wachsmotte *Galleria mellonella* L. und den kleinen Beutenkäfer *Aethina tumida* Murray bedroht. Neben *A.*

tumida, der die Bienenzucht in den USA und Australien bedroht, stellt *Varroa destructor* für die Bienenhaltung weltweit das größte Problem dar.

Die genaue Ursache für das Absterben von mit Varroa infizierten Bienenvölkern wird kontrovers diskutiert. Einige Forscher gewichten den Energieverlust durch Hämolymphverlust als gering. Diese Autoren mutmaßen, dass für den Kolonietod nicht die Milben selbst, sondern die durch sie übertragenen Pathogene verantwortlich sind. Bisher gab es zur Menge der Hämolymphe und der Energie, die der Bienenbrut durch die Milben tatsächlich verloren gehen, keine exakten Daten. In dieser Arbeit konnte erstmals der genaue Nährstoffbedarf der Milben mittels Hungerversuche- und kalorimetrischer Experimente demonstriert werden: Die Milben verbrauchten bis zu 28% der nicht regenerierbaren Reserven der verdeckelten Bienenbrut. Diese verlorenen Energiereserven führen zu deformierten, unterentwickelten und untergewichtigen adulten Bienen - dem typischen „Milben-Syndrom". Obwohl nicht ausgeschlossen werden kann, dass Viren und Bakterien an diesen Missbildungen mitwirken, sind die Milben selbst sicherlich als die Hauptursache anzusehen.

Es wurden zahlreiche Versuche unternommen, milbenresistente Bienenvölker zu züchten bzw. die Bienen über das Verständnis des Wirtsfindemechanismus von Varroa und der Manipulation der entsprechenden Parameter zu schützen. Diese Ansätze sind jedoch sehr zeitintensiv. Heute stehen mit verschiedenen biotechnischen und chemischen Techniken kurzfristiger wirksame Kontrollmethoden gegen die Milben zur Verfügung. Die biotechnischen Methoden sind jedoch arbeitsaufwendig und nicht sehr wirksam. Eine effektivere Milbenkontrolle kann durch den Einsatz bestimmter Akarizide erreicht werden, die jedoch häufig toxisch sind und Rückstände in Honig, Wachs und Propolis (Kittharz) hinterlassen. Außerdem führen viele der herkömmlichen Akarizide zu resistenten Milbenpopulationen. Die aktuelle Forschung konzentriert sich daher auf die Nutzung organischer Säuren und sekundärer Metabolite, darunter ätherische Öle aus Kräutern und Gewürzen, als Milbenbekämpfungsmittel. Angestrebt wird eine Mischung aus mehreren Komponenten mit verschiedenen Wirkweisen, da hierbei die Herausbildung von Resistenzen weniger wahrscheinlich ist. Eine derartige Mischung natürlicher Substanzen ist Propolis.

Obwohl es im Bienenstock in ausreichender Menge vorhanden ist und daher dem Imker keine Kosten verursacht, ist die Eignung von Propolis zur Milbenbekämpfung bisher nicht weiter erforscht worden. Die antibakterielle und antifungale Wirkung von Propolis ist gut bekannt. In der vorliegenden Arbeit wurde die akarizide Wirkung von Propolisproben aus unterschiedlichen geographischen Gebieten untersucht.

Propolis narkotisierte und tötete die Milben. Die Intensität dieser Effekte hing von der Art des Lösungsmittels (70%iges oder 40%iges Äthanol oder Wasser), der Propoliskonzentration und der Kontaktdauer ab. In 70%igem Äthanol gelöstes Propolis war hochtoxisch und narkotisierte unabhängig von seiner Konzentration und der Kontaktzeit 100% der Milben. Bei geringeren Propoliskonzentrationen (0,5%, 1% und 2%) erwachten die meisten Milben aus der Narkose, nur wenige starben. Höhere Propoliskonzentrationen (5% und 7.5%) führten zu einer höheren Mortalitätsrate. In 40%igem Äthanol gelöstes Propolis narkotisierte und tötete die Milben weniger effektiv als in 70%igem Äthanol gelöstes Propolis. Der Effekt von in Wasser gelöstem Propolis war vernachlässigbar gering.

Daraus läßt sich folgern, dass die wasserunlöslichen Komponenten des Propolis die Milben schädigen, die wasserlöslichen Bestandteile hingegen kaum aktiv sind. Unter natürlichen Bedingungen ist Propolis im Bienenstock wegen der Wasserunlöslichkeit seiner akariziden Komponenten nicht gegen Milben wirksam. In Äthanol aufbereitetes Propolis könnte jedoch gut als Bekämpfungsmittel eingesetzt werden. Hierfür spricht auch, dass durch eine Behandlung mit 10% Propolis in 55%igem Äthanol die Milben getötet werden, die Bienen hingegen unbeeinflußt bleiben und weder narkotisiert noch aggressiv sind. Das zeigten vorangegangene (in der vorliegenden Arbeit nicht dargestellte) Versuche.

Geringere Propoliskonzentrationen führten bei den Milben zwar nicht zum Tod, reduzierten jedoch ihre Wärmeproduktionsrate signifikant. Dies ermöglicht den Einsatz auch nicht-letaler, geringerer Propoliskonzentrationen zur Milbenbekämpfung. Die geringere Effektivität einer Behandlung mit einer geringeren Konzentration kann mit einer höheren Inkubationstemperatur (z. B. 40 °C) ausgeglichen werden.

Vergleichende Untersuchungen zur akariziden Wirkung von Propolisproben aus verschiedenen geographischen Regionen zeigten nur geringe Unterschiede bezüglich der Stärke des narkotisierenden Effektes. Die Unterschiede lassen sich durch die abweichenden chemischen Zusammensetzungen der Proben erklären, die sich aus der jeweils regionaltypischen Flora der Gebiete ergeben. Die Tatsache, dass einige Proben aus verschiedenen Regionen dieselbe akarizide Wirkung zeigten, ist ein Hinweis darauf, dass Propolis von den Bienen weltweit für denselben Zweck gesammelt wird.

Neben *V. destructor* reagierte auch die Große Wachsmotte *Galleria mellonella* auf eine Propolisbehandlung mit Narkotisierung und einer erheblich reduzierten Wärmeproduktionsrate. Die Empfindlichkeit der Großen Wachsmotte gegenüber Propolis war bei den jüngeren Larvalstadien besonders hoch. Höhere Propoliskonzentrationen töteten die Larven. Geringere Konzentrationen, die Motten im siebten Larvalstadium verabreicht wurden, beeinflussten den

Verlauf der Metamorphose. Stärke und Ausprägung dieses Effektes waren von der Propoliskonzentration abhängig.

Nach der Behandlung des siebten Larvalstadiums mit subletalen Dosen entwickelten sich die Larven zwar zur Puppe, wiesen aber verminderte Wärmeproduktionsrate während der Häutung auf. Der Grad der Störung der anschließenden Metamorphose war in hohem Maße von der Propoliskonzentration abhängig. Nach einer Behandlung mit sehr schwach konzentriertem Propolis (z.B. 1% w/v) waren die Puppen zu einer vollständigen Metamorphose in der Lage. Allerdings verlief diese schneller als bei unbehandelten Kontrolle. Dies lässt vermuten, dass Propolis in geringen Konzentrationen eine Rolle als Wachstumsregulator bei Insekten spielen könnte. Die nach einer derartig verkürzten Metamorphose geschlüpften adulten Wachsmotten wiesen schwach ausgeprägte Wärmeproduktionsrate-Zeit (p-t) Kurven auf – ein Indiz für eine schlecht entwickelte Flugfähigkeit. Die verminderte Flugfähigkeit könnte z.B. auf eine Unterentwicklung der Flugmuskeln zurückzuführen sein. Bei höheren subletalen Propoliskonzentrationen (z.B. 4% w/v) entwickelten sich zwar sämtliche Wachsmotten-Larven zu Puppen, aber die pupale Metamorphose war hier vollständig gestört. Hieraus ergibt sich die Perspektive, auch subletale Propolisdosen zur Bekämpfung der Großen Wachsmotte einsetzen zu können. Die Effekte von geringen subletalen Propolisdosen auf die Entwicklung von *Galleria mellonella* sind visuell schwierig zu beurteilen, da es nach der Behandlung weder zum Absterben noch zu sichtbaren Beeinträchtigungen der Tiere kommt. Mit Hilfe der Kalorimetrie konnten in diesen Fällen jedoch die verringerten Stoffwechsel- und Flugaktivitäten aufgezeigt werden.

Die Überprüfung der insektiziden bzw. akariziden Wirkung von Propolis mit Hilfe der Kalorimetrie ist den herkömmlichen Methoden stark überlegen. Die einzige Einschränkung der kalorimetrischen Untersuchung besteht darin, dass es wegen der benötigten Zeit für die thermische Stabilisierung des Kalorimeters nicht möglich war, die Wärmeproduktionsrate innerhalb der ersten 30 bis 45 Minuten nach der Propolisbehandlung zu beobachten. Um dieses Problem zu lösen, kann die Kalorimetrie mit Infrarot-CO_2- oder O_2-Sensoren gekoppelt werden, die den Kohlendioxid- bzw. Sauerstoffgehalt der Luft messen.

Die antimikrobielle Wirkung von Propolisproben aus verschiedenen geographischen Gebieten wurde anhand einiger Parameter wie der minimalen inhibitorischen Konzentration, der minimalen bakteriziden Konzentration, dem Durchmesser der Hemmhöfe sowie einigen Parametern der p-t-Kurven bestimmt. Die wichtigsten benutzten kalorimetrischen Parameter waren die Stärke des Abfallens der Kurven nach der jeweiligen Behandlung, die für den Wiederanstieg benötigte Zeit und die Höhe der p-t Spitze nach der Behandlung.

Der äthanolische Propolisextrakt (EEP) war gegenüber allen getesteten Mikroorganismen höchst effektiv. Die Tendenz der Dosis-Effekt-Kurve von Konzentration gegen Effekt auf die *p-t*-Kurve war bei allen Propolisextrakten ähnlich, aber die gleiche Wirkung wurde von unterschiedlichen Konzentrationen erreicht. Um die gleiche Wirkung auf die *p-t* Kurve zu erreichen, wurde die Konzentration der drei Extrakten wie folgt benötigt: EEP < PV (flüchtige Bestandteile von Propolis) < WEP (wässrige Propolisextrakte). Bei einigen Propolisproben gab es keine nennbaren Unterschiede in der antimikrobiellen Wirkung zwischen EEP und PV. Die höhere antimikrobielle Wirkung von PV könnte bedeuten, dass diese Propolisfraktion eventuell auch unter natürlichen Bedingungen eine Rolle in der Reduzierung der bakteriellen Flora innerhalb des Bienenstockes spielt. Die WEP hatten die geringsten antimikrobiellen Effekte, und deren Wirkungen waren nur bei den höchsten Propoliskonzentrationen zu beobachten. Dies verdeutlicht, dass die WEP *in situ* keine gewichtige Rolle in der Abwehr von Parasiten und Pathogenen spielen können.

Der Grund für die wesentlich höhere antimikrobielle Effektivität von EEP gegenüber den WEP und PV besteht darin, dass mit Äthanol sowohl die wasser- als auch die äthanollöslichen sowie die flüchtigen Komponenten des Propolis extrahiert werden. Durch diesen gegenüber WEP und PV wesentlich erhöhten Ertrag an wirksamen Komponenten sollte EEP das Mittel der Wahl in der Bekämpfung von Parasiten und Pathogenen im Bienenstock sein. In Situationen, in denen der Einsatz alkoholischer Extrakte nicht in Frage kommt (z.B. aus religiösen Gründen), könnte die Verwendung hochkonzentrierter WEP eine Alternative sein.

Grampositive Bakterien waren generell sensibler gegenüber Propolis als gramnegative Bakterien und Pilze. Unter den zwei getesteten gramnegativen Bakterien war *E. coli* besonders resistent. *P. syringae* reagierte zwar sensitiv auf die meisten Behandlungen, im Vergleich zu den grampositiven Bakterien wurden jedoch wesentlich höhere Propoliskonzentrationen benötigt. Die geringere Empfindlichkeit der gramnegativen Bakterien läßt sich mit der Impermeabilität ihrer äußeren Membran gegenüber antibakteriellen Agenzien erklären. Bei *E. coli* könnte eine so genannte „multiple-drug-Pumpe" für die besonders hohe Resistenz verantwortlich sein, wie sie bereits für verschiedene Stämme dieses Bakteriums nachgewiesen worden ist.

Die drei Schimmelpilze waren gegenüber Propolis unempfindlicher als die Hefe *Saccharomyces cerevisiae*, die insbesondere bei höheren Propoliskonzentrationen deutlich größere Hemmhöfe aufwies. Abgesehen vom relativ höheren Wert der minimalen inhibitorischen Konzentration ist *S. cerevisiae* in ihrer Sensitivität gegenüber konzentriertem Propolis den grampositiven Bakterien vergleichbar.

Die Untersuchung der antifungalen Aktivität von Propolis durch Petrischalen-Experimente war nur eingeschränkt möglich, da sich Propolis schlecht in der Agarschicht löst. Dies wurde durch die gesättigte Dosis-Effekt-Kurve von Hemmhofdurchmesser gegen Propoliskonzentration bei allen Schimmelpilze deutlich. Die verzögerte Propolisdiffusion war bei den bakteriellen Petrischalen-Experimenten auch zu beobachten. Hier war sie jedoch weniger stark ausgeprägt, was vermutlich an der weniger dichten Konsistenz des Nährmediums lag. Das Problem könnte durch die Verwendung eines flüssigen Nährmediums anstelle des Agars gelöst werden. Dies wäre allerdings nur für Experimente mit Bakterien- und Hefekulturen möglich.

Wegen der Eigenfarbe von Propolis und der Emulsionsbildung nach Zugabe von Propolis zu einem Flüssigmedium ist es unmöglich, die Wirkung von Propolis spektrophotometrisch zu untersuchen. Kalorimetrische Methoden werden hingegen von diesen Phänomenen nicht beeinflußt. Darüber hinaus ist die Kalorimetrie empfindlicher als die Petrischalen-Experimente und kann im Gegensatz zu letzteren bereits bei sehr geringen Propoliskonzentrationen antibakterielle Effekte nachweisen. Ein weiterer Vorteil der Kalorimetrie besteht darin, dass die Stoffwechselraten der Bakterien kontinuierlich verfolgt werden können. Im Gegensatz dazu spiegeln die Petrischalen-Experimente nur die Endergebnisse einer 24 - 48-stündigen oder längeren Zeitspanne wieder (je nach Dauer der Inkubationszeit).

Allerdings ist die Kalorimetrie auch mit Nachteilen verbunden. Die Schlauchverbindung zwischen dem Fermenter und der Kalorimeter spirale hat zur Folge, dass die Sauerstoffkonzentration in der Schlauchleitung bei höheren Zelldichten drastisch abfällt und Artefakte in den p-t Kurven verursacht. Erfolgt die Behandlung der Versuchsorganismen vor dem Erreichen der p-t Spitze, spiegelt die Kurve jedoch die Vorgänge im Fermenter genau wider und ist eine besonders empfindliche Darstellungsweise der antimikrobiellen Wirkung von biologischen Bekämpfungsmitteln wie Propolis.

9.0 General Summary / Allgemeine Zusammenfassung

9.1 General summary

The ideal conditions in the beehive interior expose the colony to several parasites, pests, and pathogens. Moreover, the high density of individuals in honeybee colonies facilitates the transmission of parasites and pathogens in the hive. Presently, beekeeping with the western honeybee *Apis mellifera* L. is endangered by the parasitic mite *Varroa destructor* (Anderson and Trueman). In addition, pests such as *Galleria mellonella*, and bacterial, fungal, and viral pathogens also remain a common problem.

The energy and nutritional demands of the parasitic lifestyle of Varroa mites have been demonstrated in this thesis, for the first time, through mite starvation and calorimetric experiments. The mites suck up to 28% of the non-replenishable reserve food of the capped brood that otherwise would have been consumed by the pupa during metamorphosis. This in turn leads to the emergence of crippled bees, and can be aggravated by viral, bacterial, and fungal infections.

The antivarroa action of propolis (bee glue) has been investigated for the first time calorimetrically, respirometrically, and using the Petridish bioassay method. Propolis samples collected from different geographic origins showed Varroa narcotizing and varroacidal effects. The length of narcosis and mortality rate of mites depended on the solvent of extraction (70% or 40% ethanol, or water), concentration of propolis, and contact time. Propolis extracted in 70% ethanol has been found to be highly toxic, resulting in the death of 80% to 100% mites regardless of the contact time and concentration of propolis. The treatment with 40% ethanol-extracted propolis was less effective, followed by the water extracted propolis.

The insecticidal action of propolis has been demonstrated by dipping the different larval stages of the greater wax moth *Galleria mellonella* in ethanol-extracted propolis for 30 s. The treatment with propolis narcotized the larvae, and reduced their metabolic rates remarkably. Higher concentrations of propolis such as 10% (w/v) caused larval mortality, whereas lower ones (1% and 2%) displayed insect growth regulator actions by shortening the length of pupal metamorphosis. A 4% w/v propolis resulted in the abortion of metamorphosis.

The antimicrobial actions of different extracts of propolis from various geographic origins were compared by using parameters such as the minimal inhibitory concentrations (MIC), minimal bactericidal concentrations (MBC), diameter of the inhibition zone, and several parameters of the heat production rate vs. time (*p-t*) curves. For all propolis samples tested, the strength of antimicrobial activity decreased in the order of ethanol-extracted propolis (EEP), propolis volatiles (PV), and water-extracted propolis (WEP). For some propolis samples, however, there was no remarkable difference in the strength of antimicrobial activities between EEP and PV. The extraction of propolis with ethanol procures all water soluble, ethanol soluble, and the volatile components of propolis making EEP superior to the other two extracts qualitatively and/or quantitatively.

The Gram positive bacteria were more sensitive to propolis treatments than the Gram negative ones and the fungi. The lower sensitivity of Gram negative bacteria can be accounted for by the impermeability of their outer membrane to antibacterial agents. Filamentous fungi were less sensitive to propolis treatment than yeasts, the latter showing higher diameters of inhibition zones especially at higher propolis concentrations.

The antimicrobial, insecticidal, and acaricidal actions of propolis illustrated here show that propolis can be used in the control of parasites, pests and pathogens of the honeybees.

9.2 Allgemeine Zusammenfassung

Bienenvölker sind im Inneren der Beute dem Kontakt mit verschiedenen Parasiten, Seuchen- und Krankheitserregern ausgesetzt. Der enge Kontakt zwischen den Bienen erhöht die Infektionsgefahr innerhalb der Kolonie. Zur Zeit wird die Bienenhaltung der westlichen Honigbiene *Apis mellifera* L. am stärksten durch die parasitisch lebende Milbe *Varroa destructor* (Anderson und Trueman) bedroht. Zusätzlich bereiten den Imkern Krankheiten wie der Befall durch die Große Wachsmotte *Galleria mellonella* und durch Bakterien, Pilze und Viren Probleme.

In dieser Arbeit wurde zum ersten Mal die genaue Energiemenge bestimmt, die die Varroa-Milbe den Bienen durch ihren parasitischen Lebensstil entzieht. Der Energiebedarf wurde durch das Aushungern der Milben und kalorimetrische Versuche ermittelt. Die Milben verbrauchen bis zu 28% der Futterreserven der verdeckelten Brut. Diese Reserven werden nicht wieder ergänzt, so dass die Nahrung den Puppen während der Metamorphose fehlt. Daher schlüpfen aus den befallenen Zellen Bienen, die unter dem typischen „Milben-Syndrom" leiden und aufgrund ihrer geschwächten Konstitution leicht das Opfer von Infektionen werden.

Die Wirkung von Propolis gegen Varroa-Milben ist in dieser Arbeit zum ersten Mal mit kalorimetrischen, respirometrischen und Petrischalen Experimenten untersucht worden. Propolis, das aus verschiedenen Regionen und Kontinenten stammte, narkotisierte und tötete die Milben. Die Länge der Narkose und die Mortalitätsrate der Milbe hing von der Wahl des Lösungsmittels, von der Propolis-Konzentration und von der Dauer des Kontaktes ab. Propolis, dass in 70%-igem Äthanol gelöst wurde, wirkte hochgradig giftig auf die Milben. Es tötete 80 bis 100% der Milben, unabhängig von der Dauer des Kontaktes und der Konzentration. Die Behandlung mit

Propolis, dass in 40%-igem Äthanol gelöst wurde, war weniger effektiv, gefolgt von in Wasser gelöstem Propolis.

Die insektizide Wirkung von Propolis wurde gezeigt, indem die verschiedenen Entwicklungsstadien der Großen Wachsmotte *Galleria mellonella* für 30 Sekunden in äthanol-gelöstes Propolis getaucht wurden. Diese Behandlung betäubte die Larven und reduzierte ihre Stoffwechselrate beträchtlich. Höhere Propolis-Konzentrationen (10% w/v) führten zum Tod der Larven, während niedrigere Konzentrationen (1% und 2%) sich regulativ auf das Larvenwachstum auswirkten und zur Verkürzung der Metamorphose führten. Eine 4%-ige Lösung führte zum Abbruch der Metamorphose.

Die antimikrobielle Wirkung der Propolisextrakte aus verschiedenen geographischen Regionen wurde anhand von Parametern wie der Minimalen inhibitorischen Konzentration (MIK), der Minimalen bakteriziden Konzentration (MBK), dem Durchmesser der Hemmhöfe und einigen Parametern der Wärmeproduktionsrate / Zeit (*p-t*) Kurven verglichen. Bei allen untersuchten Proben zeigte das in Äthanol gelöste Propolis (EEP) die stärkste antimikrobielle Wirkung, gefolgt von den flüchtigen Substanzen im Propolis (PV) und in Wasser gelöstem Propolis (WEP). Bei einigen Proben gab es nur geringe Unterschiede zwischen PV und EEP.

Löst man Propolis in Äthanol, enthält dieser Ansatz alle wasserlöslichen, alkohollöslichen und flüchtigen Komponenten des Propolis. Dadurch ist in Äthanol gelöstes Propolis den beiden anderen Formen qualitativ und quantitativ überlegen.

Grampositive Bakterien verhielten sich gegenüber Propolis grundsätzlich empfindlicher als gramnegative Bakterien und als Pilze. Die geringere Empfindlichkeit der gramnegativen Bakterien könnte darin begründet liegen, dass ihre äußere Membran für antibakterielle Substanzen undurchdringlich ist. Schimmelpilze waren Propolis gegenüber unempfindlicher als Hefen. Bei letzteren hatten die Hemmhöfe größere Durchmesser; insbesondere bei höheren Propolis-Konzentrationen.

Die hier beschriebenen antimikrobiellen, insektiziden und akariziden Wirkungen von Propolis zeigen, dass Propolis als Mittel gegen die Parasiten, Seuchen und Krankheitserreger der Honigbiene eingesetzt werden kann.

10.0 Outlook

This thesis is a pioneering research in the field of application of propolis for the control of the honeybee parasitic mite *Varroa destructor* and the honeybee pest *Galleria mellonella*. Therefore, a lot has to be done before propolis can be applied in the beehive environment. Some of the most important tasks that have to be accomplished are summarized below.

A preliminary *in vivo* experiment by spraying bees with mites on their surface displayed that propolis killed the mites but did not affect the bees. This was not, however, exhaustive, and investigations have to be conducted on the behaviour of the bees by applying propolis in laboratory and field experiments.

In addition to the use of the crude ethanol extracts of propolis, it may be desirable to fractionate it and screen for its active fractions. The use of only the active fractions of propolis could reduce the burden of unnecessary chemicals on the bees and in the hive products.

Chemical analyses of the effective antivarroa fractions of propolis have to be done, and the chemicals responsible for the antivarroa action have to be identified. Identification of the chemical make-up of the effective fractions of propolis may help in the search for such fractions from abundantly available sources, since the availability of propolis is usually limited.

Application of propolis in the beehive could be done by spraying or dripping a solution of it; these methods have to be optimized, and the proper application instruments have to be developed.

The antivarroa actions of propolis investigated in this thesis were its Varroa narcotizing and varroacidal effects. It is, however, possible that the volatile components of propolis could affect the mites by interfering with their chemical cue and orientation; this remains to be investigated. The effect of some essential oils on orientation of Varroa mites has already been demonstrated by several researchers.

In order to use propolis as an agent to control *Galleria mellonella,* methods of application in weak colonies, and in honeycomb storehouses have to be developed. The identification and use of only the effective fractions of propolis is desirable, as such fractions reduce the chemical burden in the beehive.

In order to elucidate the exact mechanism of action of propolis against bacteria and fungi, detailed investigations have to be conducted about the effect of propolis on cell division and associated events, protein synthesis, cell membrane function and integrity, cell wall function and integrity, and so on. Some of these can be performed by transmission electron microscopic (TEM) technique.

Practice oriented investigations about the potential applications of propolis against saprophytic fungi that destroy paper products in libraries and archives, and degrade leather have to be carried out.

11.0 References

Aga H., Shibuya T., Sugimoto T., Kurimoto M., Nakajima S.H. (1994). Isolation and identification of antimicrobial compounds in Brazilian propolis. Biosc. Biotech. Biochem. 58, 945-946.

Amoros M., Sauvager F., Girre L., Cormier M. (1992). *In vitro* antiviral activity of propolis. Apidologie 23, 231-240.

Amos T. G., Williams P., Du Guesclin P. B., Schwarz M. (1974). Compounds related to Juvenile Hormone: activity of selected terpenoids on *Tribolium castaneum* and *T. confusum*. J. Econ. Entomol. 67, 474-476.

Amrin J., Noel B., Mallow H., Stasny T., Skidmore R. (1996) Results of research: using essential oils for honey bee mite control. http://www.wvu.edu/agexten /varroa.htm.

Anderson D. and Trueman J.W.H. (2000) *Varroa jacobsoni* (Acari: Varroidae) is more than one species. *Exper. App. Acarol.* 24, 165-189.

Ayala F., Lembo G., Nappa P., Balato N. (1985) Contact dermatitis from propolis. Cont. Dermat. 12, 181–182.

Baier K. (1969) Die Wasserdampfsperre in der Beute. Bienenpflege, Weinsberg 7, 143-147.

Bailey L. and Ball B. V. (1991) Honeybee pathology. London, Academic press limited, 2nd ed. pp. 193.

Ball B. V. (1983) Der Zusammenhang zwischen *Varroa jacobsoni* und Viruserkrankung der Honigbiene. Allg. Deutsche Imker 17, 177-179.

Ball B. V. (1994) Host-parasite-pathogen interactions. In: New perspectives on *Varroa*, Matheson A. (ed.) Cardiff, UK, IBRA. pp 5-11.

Ball B. V. (1996) Honeybee viruses: a cause for concern? Bee World 77, 117-119.

Ball B.V. and Allen M. F. (1986) The incidence of acute paralysis virus in honeybee colonies infested with the parasitic mite *Varroa jacobsoni*. In: Fundamental and applied aspects on invertebrate pathology, Samson R. A., Vlak J. M., Peters D. (eds.) 4th Int Colloq. Invertebr. Pathol., p 151.

Bankova V., Boudourova-Krasteva G., Sforcin J. M., Frete X., Kujumgiev A., Maimoni-Rodella R., Popov S. (1999) Phytochemical evidence for the plant origin of Brazilian propolis from Sa˜o Paulo State. Z. Naturforsch. 54c, 401-405.

Bankova V., Christov R., Popov S. (1994) Volatile constituents of propolis. Z. Naturforsch. 49c, 6-10.

Bankova V., Dyulgerov A., Popov S. (1992) Propolis produced in Bulgaria and Mongolia: phenolic compounds and plant origin. Apidologie 23, 79-85.

Bankova V., Marcucci M.C., Simova S., Nikolova N., Kujumgiev A., Popov S. (1996) Antibacterial diterpenic acids from Brazilian propolis. Z. Naturforsch. 51c, 277–280.

Bankova V., Popova M., Bogdanov S., Sabatini A.-G. (2002) Chemical composition of European propolis: expected and unexpected results. Z. Naturforsch. 57c, 530-533.

Bankova V.S., De Castro S. L., Marcucci M. C. (2000) Propolis: recent advances in chemistry and plant origin. Apidologie 31, 3-15.

Banskota A.H., Tezuka Y., Prasain J.K., Matsushige K., Saiki I., Kadota Sh. (1998) Chemical constituents of Brazilian propolis and their cytotoxic activities. J. Nat. Prod. 61, 896–900.

Baxter J., Eischen F., Pettis J., Wilson W.T., Shimanuki H. (1998) Detection of fluvalinate-resistant *Varroa* mites in U.S. honey bees. Am. Bee J. 138, 291.

Beetsma J., Boot W.J., Callis J.N.M. (1999) Invasion behaviour of *Varroa jacobsoni* Oud. from bees to brood cells. Apidologie 30, 125-140.

Beezer A.E., Hills A.K., O'Neill M.A.A., Morris A.C., Kierstan K.T.E., Deal R.M., Waters L.J., Hadgraft J., Mitchell J.C., Connor J.A., Orchard J.E., Willson R.J., Hofelich T.C., Beaudin J., Wolf G., Baitalow F., Gaisford S., Lane R.A., Buckton G., Phipps M.A., Winneke R.A., Schmitt E.A., Hansen L.D., O'Sullivan D., Parmar M.K. (2001) The imidazole catalysed hydrolysis of triacetin: an inter- and intra-laboratory development of a test reaction for isothermal heat conduction microcalorimeters used for determination of both thermodynamic and kinetic parameters. Thermochim. Acta 380, 13-17.

Benedek P. (1985). Economic importance of honeybee pollination of crops at the national level in Hungary. Proceedings of the 29th. international beekeeping congress, 25-31 August 1983, Budapest, Hungary, *Apimondia,* 286-289.

Bjorkner B. E. (1994). Industrial airborne dermatoses. Dermat. Clin. 12, 501-509.

Boecking O. and Ritter W. (1993) Grooming and removal behaviour of *Apis mellifera intermissa* in Tunisia against *Varroa jacobsoni*. J. Apic. Res. 32, 127-134.

Boecking O. and Spivak M. (1999) Behavioural defense of honey bees against *Varroa jacobsoni* Oud. Apidologie 30, 141-158.

Bogdanov S. (1999) Propolis: Harvest, composition and quality. Swiss centre of Bee research http://www.apis.admin.ch/deutsch/pdf/Bienenprodukte/Propolis_d.pdf

Bogdanov S., Kolchenmann V., Imdorf A. (1998) Acaricide residues in some bee products. J. Apic. Res. 37, 57-67.

Bollhalder F. (1999) Trichogramma for wax moth control. Am. Bee J. 136, 711-712.

Bonvehi S.J. and Coll V. (2000) Study on propolis quality from China and Uruguay. Z. Naturforsch. 55c, 778-784.

Bonvehi S.J., Coll V. F., Jorda E. R. (1994) The composition, active components and bacteriostatic activity of propolis on dietetics. J. Am. Oil Chem. Soc. 71, 529-532.

Boot J. W., Calis J. N. M., Beetsma J. (1991) Invasion of *Varroa* mites into honeybee brood cells. When do brood cells attract *Varroa* mites? Proceedings of the section of experimental and applied entomology of the Netherlands entomological society (N.E.V) 2, 154-156.

Borchert A. (1966) Die Krankheiten und Schädlinge der Honigbiene. Hirzel Verlag, Leipzig. 3rd edn. pp. 123.

Bowen-Walker P. L. and Gunn A. (2001) The effect of the ectoparasitic mite, *Varroa destructor* on adult worker honeybee (*Apis mellifera*) emergence weights, water, protein, carbohydrate, and lipid levels. Entomologia Experimentalis et Applicata 101, 207–217.

Briggner L.E. and Wadsö I. (1991) Test and calibration processes for microcalorimeters, with special reference to heat conduction instruments used with aqueous systems. J. Biochem. Biophys. Meth. 22, 101-111.

Brødsgaard C. J. and Hansen H. (1994) An example of integrated biotechnical and soft chemical control of *Varroa* in a Danish apiary. In: New perspectives on *Varroa,* Matheson A. (ed.) Cardiff, UK: IBRA. pp 101-105.

Brown M.R. and Williams P. (1985) The influence of environment on envelope properties affecting survival of bacteria in infections. Annu. Rev. Microbiol. 39, 527-556.

Buchler R. (1994) *Varroa* tolerance in honeybees-occurrence, characters and breeding. In: New perspectives on *Varroa*. International Bee Research Association, Matheson A. (ed.) Cardiff, UK, pp.12-23.

Burdock G.A. (1998) Review of the biological properties and toxicity of bee propolis. Food and Chem. Toxicol. 36, 347-363.

Calis J.N.M., Boot W. J., Beetsma J., van den Eijnde J.H.P.M., de Ruijter A., van der Steen. J. J. (1998) Control of *Varroa* by combining trapping in honey bee worker brood with formic acid treatment of the capped brood outside the colony: putting knowledge on brood cell invasion into practice. J. Apic. Res. 37, 205-215.

Caron D.M (1992) The wax moth. Am. Bee J. 132, 647-649.

Charrière J.D. and Imdorf A. (1997) Protection of honeycombs from moth damage. Swiss Bee Research center, Federal Dairy and Research Station, Liebefel, Bern. Commun. No. 24.

Cheng P. and Wong G. (1996) Honeybee propolis: prospects in medicine. Bee World 77, 8-15.

Cizmarik J. and Trupl J. (1975) The action of propolis on saccharomycetes. Pharmazie 30, 406-407.

Colombo M. Lodesani M., Spreafico M. (1993) Resistance of *Varroa jacobsoni* to fluvalinate, preliminary results of investigations conducted in Lombardy. Ape nostra Amica 15, 12-15.

Corrêa-Marques M.H. and De Jong D. (1998) Uncapping of worker bee brood, a component of the hygienic behaviour of Africanized honey bees against the mite *Varroa jacobsoni* Oud. *Apidologie 29*, 283-289.

Crane E. (1979) Fresh news on the *Varroa* mite. Bee world 60, 8.

Crane E. (1990) Bees and beekeeping: Science, Practice and World Resources. Cornstock Publ., Ithaca, NY., USA. pp. 593.

Croxton F.E., Cowden D.J., Klein S. (1967) Applied general statistics. 3rd ed. Prentice-Hall. Englewood Cliffs, N.J., pp.754.

De Castro S. L. and Higashi K. O. (1995) Effect of different formulations of propolis on mice infected with *Trypanosoma cruzi*. J. Ethnopharm. 46, 55-58.

De Jong D. (1996) Africanized honey bees in Brazil, forty years of adaptation and success. Bee World 77, 67-70.

De Jong D. (1997) Mites: *Varroa* and other parasites of brood. In: Honey bee pests predators and disease, Morse R. M, Flottum P. K., (eds.), Medina, OH Root 3rd ed. 2, 281-327

De Jong D., and De Jong P. H. (1983) Longevity of Africanized honeybees (Hymnoptera: Apidae) infested by *Varroa jacobsoni* (Parasitiformes: Varroidae). J. Econ. Entomol. 76, 766-768.

De Jong D., De Jong P. H., Gonçalves L.S. (1982) Weight loss and other damage to developing worker honey bees from infestation with *Varroa jacobsoni*. J. Apicult. Res. 21, 165-167.

De Jong D., Gonçalves L. S., Morse R. A. (1984) Dependence on climate of the virulence of *Varroa jacobsoni*. Bee World 65, 117-121.

De Ruijter A. and Pappas N. (1983) Karyotype and sex determination of *Varroa jacobsoni* Oud. In: *Varroa jacobsoni* Oud. affecting honeybees: present status and needs, Cavalloro R. (ed.), A.A. Balkema, Rotterdam pp.41-44.

Delfinado Baker M. (1988) The tracheal mite of honeybees: a crisis in beekeeping. In: Africanized honey bees and bee mites, Needham G., Page R., Delfinado-Baker M., Bowman C. (eds.) Halsted press, Chiehester, England, pp 493-497.

Denyer S.P. and Stewart G.S.A.B. (1998) Mechanisms of action of disinfectants. Internl. Biodeterior. and Biodegrad. 41, 261–268.

Digrak M., Yilmaz O., Ozscelik S. (1995) *In vitro* antimicrobial effect of propolis in Elazig region. Turk. J. Biol. 19, 249-257.

Dobrowolski J. W., Vohora S. B., Sharma K., Shah S. A., Naqvi S. A. H., Dandiya P. C. (1991). Antibacterial, antifungal, antiamoebic, antiinflammatory and antipyretic studies on propolis bee products. J. Ethnopharm. 35, 77-82.

Dobrowolski J.W., Vohora S.B., Sharma K., Shah S.A., Naqvi S.A.H., Dandiya P.C. (1991) Antibacterial, antifungal, antiamoebic, antiinflammatory and antipyretic studies on propolis bee products. J. Ethnopharm. 35, 77-82.

Donzé G. and Guerin P.M. (1997) Time – activity budgets and space structuring by the different life stages of *Varroa jacobsoni* in capped brood of the honey bee *Apis mellifea.* J. Insect. Behav. 10, 371-393.

Donzé G., Schnyder-Candrian S., Bogdanov S. Diehl P. A., Guerin P. M. (1998) Aliphatic alcohols and aldehydes of the honeybee cocoon induce arrestment behaviour in *Varoa jacobsoni* (Acari: Mesostigmata), an ectoparasite of *Apis mellifera.* Arch. Insect Biochem. Phsiol. 37, 129-145.

Downey D.L., Higo T.T., Winston M.L. (2000) Single and dual parasitic mite infestations on the honey bees, *Apis mellifera* L. Insectes soc. 47, 171-176.

Droege G. (1989) Das Imkerbuch. VEB Deutscher Landwirtschaftsverlag. Berlin, Melsungen, pp. 220.

Eischen F.A. (1995) *Varroa* resistance to Fluvalinate. Am.Bee J. 135, 815-816

Ellis Jr. J. D. (2001) The future of *Varroa* control: integrating current treatments with the latest advancements. Am. Bee J. 141, 127-130.

Ellis Jr. J. D., Delaplane K. S., Hood W. M. (2001) Efficacy of bottom screen device, Apistan [TM] and ApilifeVAR [TM] in controlling *Varroa destructor.* Am. Bee J 141, 813-816.

Elzen P. J., Eischen F. A, Baxter J. B., Pettis J., Elzen G. W., Wilson W. T. (1998) Fluvalinate resistance in *Varroa jacobsoni* from several geographic locations. Am. Bee J. 138, 674-676.

Elzen P.J., Baxter J.R., Spivak M., Wilson W.T. (2000) Control of *Varroa jacobsoni* Oud. resistant to fluvalinate and amitraz using coumaphos. Apidologie 31, 437-441.

Engels W. (1994) *Varroa* control by hypothermia. In: New perspectives on *Varroa,* Matheson A. (ed.), Cardiff, UK, IBRA. pp 115-119.

Faye I. and Wyatt GR. (1980) The synthesis of antibacterial proteins in isolated fat body from cecropia silk moth pupae. Experientia 36, 1325-1326.

Finley J., Camazine S., Frazier M. (1997) The epidemic of honey bee colony losses during the 1995-1996 season. Am. Bee J. 136, 805-808.

Folch, J., Lees M., Sloanestanley G.H. (1957) A simple method for the isolation and purification of total lipids from animal tissues. J. Biol.Chem. 226, 497-509.

Fraser H. (1997) The effect of different conspecific male sex pheromone component ratios on the behaviour of the female greater wax moth. MSc thesis University of Guelph, Ontario, Canada, pp.102.

Frazier M. (2000) *Varroa* mites. MAAREC Publication (The Pennsylvania State University) 4.7, 1-4.

Fries I. (1991) Treatment of sealed honey bee brood with formic acid for control of *Varroa jacobsoni*. Am. Bee J. 131, 313-314.

Fries I. and Hansen H. (1993) Biotechnical control of *Varroa* mites in cold climates. Am. Bee J. 133, 435-438.

Fries I., Camazine S., Sneyd J. (1994). Population dynamics of *Varroa jacobsoni*: A model and a review. Bee World 75, 5-28.

Fuchs S. (1990) Preference for drone brood cells by *Varroa jacobsoni* Oud. in colonies of *Apis mellifera carnica*, Apidologie 21, 193-199.

Fuchs S. and Müller K. (1988) Invasion of honeybee brood cells by *Varroa jacobsoni* in relation to the age of the larvae. In: European research on varroatosis control, Cavalloro R. (ed.), A.A. Balkema, Rotterdam, Holland, pp 77-79.

Garcia-Viguera C., Ferreres F., Tomas-Barberan F.A. (1993) Study of Canadian Propolis by GC-MS and HPLC. Z. Naturforsch. 48c, 731–735.

Garedew A. and Lamprecht I. (1997) Microcalorimetric investigations on the influence of propolis on the bacterium *Micrococcus luteus*. Thermochim. Acta 290, 155-166.

Garedew A., Schmolz E., Lamprecht I. (2003) Microcalorimetric investigation on the antimicrobial activity of honey of the stingless bee *Trigona spp.* and comparison of some parameters with those obtained with standard methods. (in press: Thermochim. Acta).

Ghisalberti E.L. (1979) Propolis: a review. Bee world 60, 58-84

Ghisalberti E.L., Jefferies P.R., Lanteri R., Matisons J. (1978) Constituents of propolis. Experientia 34, 157–158.

Gliński Z. and Jarosz J. (1984) Alterations in hemolymph proteins of drone honeybee larvae parasitized by *Varroa jacobsoni*. Apidologie 15, 329-337.

Gliński Z. and Jarosz J. (1988) Deleterious effects of *Varroa jacobsoni* on the honeybee. Apiacta 23, 42-52.

Gliński Z. and Jarosz J. (1990a) Microorganisms associated fortuitously with *Varroa jacobsoni* mite. Microbios 62, 59-68.

Gliński Z. and Jarosz J. (1990b) *Serratia marcescens,* artificially contaminating brood and worker honeybees, contaminates the *Varroa jacobsoni* mite. J. Apicult. Res. 29, 107-111.

Gliński Z. and Jarosz J. (1992) *Varroa jacobsoni* as a carrier of bacterial infections to a recipient bee host. Apidologie 23, 25-31.

Gnaiger E. (1993) Nonequilibrium thermodynamics of energy transformations. Pure and Appl. Chem. 65, 1983-2002.

Grange J.M. and Davey R.W. (1990). Antibacterial properties of propolis (bee glue). J. Roy. Soc. Med. 83, 159-160.

Greenaway W., Scaysbrook T., Whatley F.R. (1987) The analysis of bud exudate of *Populus* x *euramericana*, and of propolis, by gas chromatography-mass spectrometry, Proc. R. Soc. London B 232, 249-272.

Greenaway W., Scaysbrook T., Whatley F.R. (1990) The composition and plant origins of propolis: a review of report of work at Oxford. Bee World 71, 107-118.

Haines P.J., Reading M., Wilburn F.W. (1998) Differential thermal analysis and differential scanning calorimetry. In: Handbook of thermal analysis and calorimetry: principles and practices (Vol. 1), Brown M.E. (ed.), Elsevier Science B.V., Amsterdam, The Netherlands, pp 279-361.

Harak M., Lamprecht I., Kuusik A. (1996) Metabolic cost of ventilating movements in pupae of *Tenebrio molitor* and *Galleria mellonella* studied by direct and indirect calorimetry. Thermochim. Acta 276, 41-47.

Hargasm O. (1973) Die Milbe *Varroa jacobsoni* Oudemans bedroht die Bienenzucht in Europa. Imkerfreund 28, 316-317.

Hausen B.M., Wollenweber E., Senff H., Post B. (1987) Propolis allergy. I. Origin, properties, usage and literature review. Cont. dermat. 17, 163-170.

Heal R.D. and Parsons A.T. (2002) Novel intercellular communication system in *Escherichia coli* that confers antibiotic resistance between physically separated populations. J. Appl. Microbiol. 92, 1116-1122.

Hemminger W. and Sarge S.M. (1998) Definitions, nomenclature, terms and literature. In: Handbook of thermal analysis and calorimetry: principles and practices (Vol. 1), Brown M.E. (ed.), Elsevier Science B.V., Amsterdam, The Netherlands, pp 1-73.

Hogan D. and Kolter R. (2002) Why are bacteria refractory to antimicrobials? Curr. Opin. Microbiol. 5, 472-477

Hölzel R., Motzkus C., Lamprecht I (1994) Kinetic investigation of microbial metabolism by means of flow calorimeters. Thermochim. Acta 239, 17-32.

Hoppe H. and Ritter W. (1987) Untersuchungen zur Kombinierten Wärmetherapie gegen die Varroatose. Apidologie 18, 383-384

Hoppe H. and Ritter W. (1989) The influence of the Nasonov pheromone on the recognition of house bees and foragers by *Varroa jacobsoni*. Apidologie 19, 165-172.

Horn H. (1981) Bienen im elektrischen Feld. Apidologie 12, 101-103.

Huang Z.(2001) Mite Zapper- a new and effective method for *Varroa* mites control. Am. Bee J. 141, 730-732.

Ikeno K., Ikeno T., Miyazawa C. (1991). Effects of propolis on dental caries in rats. Caries Res. 25, 347-351.

Imdorf A., Bogdanov S., Ochoa R.I., Calderone N.W. (1999) Use of essential oils for control of *Varroa jacobsoni* Oud. in honeybee colonies. Apidologie 30, 209-228.

Infantidis M. D. (1983) Ontogenesis of the mite *Varroa jacobsoni* in worker and drone brood cells. J. Apicult. Res. 22, 200-206.

Infantidis M.D. (1988) Some aspects of *Varroa jacobsoni* entrance into honeybee (*Apis mellifera*) brood cells. Apidologie 19, 387-396.

Issa M. R. C. and Gonçalves L. (1984) Study on the preference of the acarid *Varroa jacobsoni* for drones of Africanized honeybees. In: Advances in invertebrate reproduction, Engels W. (ed.), Elsvier, Amsterdam, Holland, pp. 598.

Johansen P. and Wadsö I. (1999) An isothermal microcalorimetric titration/perfusion vessel equipped with electrodes and spectrophotometer. Thermochim. Acta 342, 19-29.

Johnson K. S., Eischen F.A., Giannasi D.E. (1994) Chemical composition of North American bee propolis and biological activity towards larvae of the greater wax moth (Lepidoptera: Pyralidae). J. Chem. Ecol. 20, 1783-1792.

Kedzia B., Grppert B., Iwaszkiewicz J. (1990) Pharmacological investigations of ethanolic extract of propolis. Phytothérapie 6, 7-10.

Kemp R.B. (1998) Nonscanning calorimetry. In: Handbook of thermal analysis and calorimetry: principles and practices (Vol. 1), Brown M.E. (ed.), Elsevier Science B.V., Amsterdam, The Netherlands, pp 577-608.

Knox D. A., Shimanuki H., Herbert E. W. (1971) Diet and the longevity of adult honeybees. J. Econ. Entomol. 64, 1415-1416.

König B. (1985) Zur Naturgeschichte der Propolis. Allg. D. Imk. Zeit. 19, 310-312.

König D. and Dustmann J.H. (1988) Baumharze, Bienen und antivirale Chemotherapie. Naturwissen. Rundschau 2, 43-53.

Kovac H. and Crailsheim K. (1988) Lifespan of *Apis mellifera carnica* Pollm. infested by *Varroa jacobsoni* Oud. in relation to season and extent of infestation. J. Apiclut. Res. 27, 230-238.

Kraus B. and Velthuis H.H.W. (1997) High humidity in the honeybee brood nest limits reproduction of the parasitic mite *Varroa jacobsoni*. Naturwissenschaften 84, 217-218.

Kraus B., Königer N., Fuchs S. (1994) Screenning of substances for their effect on *Varroa jacobsoni* attractiveness, repellence, toxicity and masking effect of etheral oils. J. Apic. Res. 33, 34-43.

Kraus B., Page R.E. (1995) Effect of *Varroa jacobsoni* (Mesostigmata, Varroidae) on feral *Apis mellifera* (Hymnoptera, Apidae) in California. Environ. Entomol. 24, 1473-1480.

Kraus B., Velthuis H. H. W., Tingek S. (1998) Temperature profiles of the brood nests of *Apis cerana* and *Apis mellifera* colonies and their relation to varroatosis. J. Apicult. Res. 37, 175-181.

Krell R. (1996) Value-Added Products from Beekeeping. FAO Agricultural Services Bulletin No. 124 Food and Agriculture Organization of the United Nations, Rome.

Krol W., Scheller S., Shenai J., Pietsz G., Czuba Z. (1993) Synergistic effect of ethanol extract of propolis and antibiotics on the growth of *Staphylococcus aureus*. Arzncim-Forsch. Drug Res. 43, 607-609.

Kubik M., Nowacki J., Michalczuk L., Pidek A., Marcinkowski J. (1995) Penetration of fluvalinate into bee-products. J. Fruit Ornamen. Plant Res.3, 13-22.

Kujumgiev A., Bankova V., Ignatova S. (1993) Antibacterial activity of propolis, some of its components and their analogues. Pharmazie 48, 785-786.

Kujumgiev A., Tsvetkova I., Serkedjieva Yu., Bankova V., Christov R., Popov S. (1999) Antibacterial, antifungal and antiviral activity of propolis of different geographic origin. J. Ethnopharmcol. 64 (3) 235-240

Kuusik A., Harak M., Hiiesaar K., Metspalu L., Tartes U. (1995) Studies on insect growth regulating (IGR) and toxic effects of *Ledum palsur* extracts on *Tenebrio molitor* pupae (Coleoptera: Tenebrionidae) using calorimetric recordings. Thermochim. Acta 251, 247-253.

Kuusik A., Metspalu L., Hiiesaar K., Koegerman A., Tartes U. (1993) Changes in muscular and respiratory patterns in the yellow meal worm *(Tenebrio molitor)* and greater wax moth *(Galleria mellonella)* pupae caused by some plant extracts, juvenile hormone analogues and pyrethroids. Proc. Estonian Acad. Sci. Biol. 42, 94-107.

Lamprecht I. (1983) Application of calorimetry to different biological fields and comparison with other methods. Boll. Soc. Natur. Napoli. 92, 515-542.

Lamprecht I. (1997) Calorimetric experiments on social insects. Thermochim. Acta 300, 213-224.

Lamprecht I. (1999) Biology. In: Chemical Thermodynamics: A 'Chemistry for the 21st century', Letcher T. M. (ed.) Monograph, Blackwell Science pp. 265.

Lamprecht I., Hemminger W., Höhne G. (1991) Calorimetry in the biological sciences. Thermochim. Acta 193, 452-471.

Lavie P. (1976) The relationship between propolis, poplar buds *(Populus spp)* and castoreum. Proc. XXV Int. Beekeeping Congr., Grenoble, 1975, Apimondia Publ. House, Bucharest, pp. 229-233.

Le Conte Y. and Cornuet J. M. (1989) Variability of the post capping stage duration of the worker brood in three different races of *Apis mellifera*. In: Present status of varroatosis in Europe and progress in the *Varroa* mites control, Commission of the European communities, Cavalloro R. (ed.), Luxemberg, pp. 171-175.

Le Conte Y., Arnold G., Trouiller J., Mason C., Chappe B. (1989) Attraction of the parasitic mite *Varroa* to the drone larvae of honey bees by simple aliphatic esters. Science 245, 638-639.

Lewis K. and Lomovskaya O. (2001) Drug efflux. In: Bacterial resistance to antimicrobials: mechanisms, genetics, medical practice and public health, Lewis K., Salyers A., Taber H., Wax R. (eds.), Marcel Dekker, Inc., New York, N.Y. pp. 61–90

Lindenfelser L.A. (1967) Antimicrobial activity of propolis. Am. Bee J 107, 90-92.

Lisowski F. (1984) Demystifying health foods. On Continuing Practice 11, 11-14.

Liu T. (1996) *Varroa* mites as carriers of honeybee chalkbrood. Am. Bee J. 136, 655.

Lodesani M., Colombo M., Spreafico M. (1995) Ineffectiveness of Apistan™ treatment against the mite *Varroa jacobsoni* Oud. in several districts of Lombardy (Italy). Apidologie 26, 67-72.

Löhr K.D., Sayyadi P., Lamprecht I. (1978) Heat production and respiration during the development of two insects. In: Thermodynamics of Biological Processes, Lamprecht I. and Zotin A.I. (eds.) de Gruyter, Berlin. pp. 197.

Lomovskaya O. and Lewis K. (1992) Emr, an *Escherichia coli* locus for multidrug resistance. Proc. Natl. Acad. Sci., USA 89, 8938–8942.

Loschiavo S. R. (1975) Tests of four synthetic growth regulators with juvenile hormone activity against seven species of stored products insects. Manit. Entomol. 9, 43-51.

Maillar J.Y. (2002) Bacterial target sites for biocide action. J. Appl. Microbiol. 92, 16S-27S.

Marcangeli J., Monetti L., Fernandez N. (1992) Malformations produced by *Varroa jacobsoni* on *Apis mellifera* in the province of Buenos Aires, Argentina. Apidologie 23, 399-402.

Marcucci M.C. (1995) Propolis: chemical composition, biological properties and therapeutic activity. Apidologie 26, 83-99.

Marcucci M.C., Rodriguez J., Ferreres F., Bankova V., Groto R., Popov S. (1998) Chemical composition of Brazilian propolis from São Paulo state. Z. Naturforsch. 53c, 117–119.

Markham K.E., Mitchel K. A., Wilkins A.L., Daldy J.A., Lu Y. (1996) HPLC and GC-MS identification of the major organic constituents in New Zealand propolis. Phytochem. 42, 205-211.

Martin S. J. (1994) Ontogenesis of the mite *Varroa jacobsoni* Oud. in worker brood of the honey bee *Apis mellifera* L. under natural conditions. Exp. Appl. Acarol. 18, 87-100.

Martin S. J., Holland K., Murray M. (1997) Non-reproduction in the honeybee mite *Varroa jacobsoni.* Exp. Appl. Acarol. 21, 539-549.

Martos I., Cossentini M., Ferreres F., Tomas-Barberan F.A. (1997) Flavonoid Composition of Tunisian honey and propolis. J. Agric. Food Chem. 54, 2824–2829.

Matheson A. (1994) The impact of *Varroa* infestation on beekeeping. In: New perspectives on *Varroa*, Matheson A. (ed.), International Bee Research Association, Cardiff, UK, pp. 27-31.

Matheson A. (1995) First documented findings of *Varroa jacobsoni* outside its presumed natural range. Apiacta 30, 1-8.

Maul V., Klepschi A., Assmann-Werthmüller U. (1988) Das Bannwabenverfahren als Element Imkerlicher Betriebsweise bei starkem Befall mit *Varroa jacobsoni* Oud. Apidologie 19, 139-154.

Maurizio A. (1954) Pollenernährung und Lebensvorgänge bei der Honigbiene (*Apis mellifera* L.). Landwirtschaftl. Jahrbuch der Schweiz 3, 115-183.

Mc Devitt D., Payne D.J., Holmes D.J., Rosenberg M. (2002) Novel targets for the future development of antibacterial agents. J. App. Microbiol. 92, 28S-34S.

Medina L.M. and Martin S.J. (1999) A comparative study of *Varroa jacobsoni* reproduction in worker cells of honeybees (*Apis mellifera*) in England and Africanized bees in Yucatan, Mexico. Exp. Appl. Acarol. 23, 659-667.

Menezes H., Bacci Jr M., Oliveira S.D., Pagnocca F.C. (1997) Antibacterial properties of propolis and products containing propolis from Brazil. Apidologie 28, 71-76.

Message D. and Gonçalves L. S. (1995) Effect of the size of worker brood cells of Africanized honey bees on infestation and reproduction of the ectoparasitic mite *Varroa jacobsoni* Oud. Apidologie 26, 381-386.

Metwally M. M. and Sehnal F. (1973). Effects of Juvenile hormone analogues on the metamorphosis of beetles *Trogoderma granarium* (Dermestidae) and *Carydon gonagra* (Bruchidae). Biol. Bull. 144, 368-382.

Metzner J., Bekeimer H., Paintz M., Schneidewind E.M. (1979) On the antimicrobial activity of propolis and propolis constituents. Pharmazie 34, 97-102.

Metzner J., Schneidewind E.M., Friedrich E. (1977) Effects of propolis and pinocembrin on yeasts. Pharmazie 32, 730.

Meyer W. (1956) Propolis bees and their activities. Bee World 37, 25-36.

Milani N. (1994) Possible presence of fluvalinate-resistant strains in *Varroa jacobsoni* in northern Italy. In: New perspectives on *Varroa*, Matheson A. (ed.) Cardiff, UK: IBRA. pp 87.

Milani N. (1995) The resistance of *Varroa jacobsoni* Oud, to pyrethroids: a laboratory assay. Apidologie 26, 415-429.

Milani N., and Della Vedova G. (1996) Determination of the LC50 in the mite *Varroa jacobsoni* of the active substances in Perizin ® and cekafix ®. Apidologie 26, 67-72.

Miyataka H., Nishiki M., Matsumoto H., Fujimoto T., Matsuka M., Satoh T. (1997) Evaluation of propolis. I. Evaluation of Brazilian and Chinese propolis by enzymatic and physico-chemical methods. Biol. Pharm. Bull. 20, 496-501.

Möbus B. (1972) The importance of propolis to honeybees. Br. Bee. J. 100, 198-199.

Monti M., Berti E., Carminati G., Cusini M. (1983) Occupational and cosmetic dermatitis from propolis. Cont. dermat. 9, 163.

Moosbeckhofer R. (1993) Wachsmotten – eine Gefahr für den Wabenvorrat. Bienenvater 6, 261-270.

Moretto G., Gonçalves L. S., De Jong D. (1993) Heritability of Africanized and European honey bee defensive behaviour against the mite *Varroa jacobsoni*. Rev. Bras. Genet. 16, 71-77.

Moretto G., Gonçalves L. S., De Jong D. (1997) Relationship between food availability and the reproductive ability of the mite *Varroa jacobsoni* in Africanized bee colonies. Am Bee J. 137, 67-69.

Moritz R. F. A. (1985) Heritability of the postcapping stage in *Apis mellifera* and its relation to varroatosis resistance. J. Heredity 76, 267-270.

Morse R.A. (1978) Arachnids: Acarina (mites and ticks). In: Honeybee pests, predators and diseases, Morse R.A. (ed.) Cornell University Press. pp. 197-209

Münstedt K. and Zygmunt M. (2001) Propolis – current and future medical uses. Am. Bee J. 141, 507-510.

Mutinelli F., Baggio A., Capolongo F., Piro R., Prandin L. (1997) A scientific note on oxalic acid by topical application for the control of varroatosis. Apidologie 28 (6), 461-462.

Nagai T., Inoue R., Inoue H., Suzuki N. (2003) Preparation and antioxidant properties of water extracts of propolis. Food Chem. 80, 29-33.

Nagy E., Papay V., Litkei G., Dinya Z. (1986) Investigation of the chemical constituents, particularly the flavonoid components, of propolis and Populigemma by the GC/MS method. Stud. Org. Chem. (Amsterdam) 23. 223-232.

National Committee for Clinical Laboratory Standards (1985) Methods for dilution antimicrobial susceptibility tests for bacteria that grow aerobically. Approved standard M7-A NCCLS, Villanova, PA.

Neumann P., Pirk C.W.W., Hepburn H.R., Solbrig A.J., Ratnieks F.L.W., Elzen P.J., Baxter J.R. (2001) Social encapsulation of beetle parasites by cape honeybee colonies (*Apis mellifera capensis* Esch). Naturwissenschaften 88, 214-216.

Neunaber E. (1995) Phytochemische und mikrobiologische Untersuchungen von Propolis verschiedener Provenienzen als Beitrag zur Kenntnis der Wirkprinzipien in Propolis. Inaugural-Dissertation Free University of Berlin, Faculty of Pharmacy, Berlin.

Nikaido H. (1998) Multiple antibiotic resistance and efflux. Curr. Opin. Microbiol. 1, 516-523.

Nikaido H. (1999). Microdermatology: cell surface in the interaction of microbes with the external world. J. Bacteriol. 181, 4–8.

Nishino T., Wecke J., Kruger D., Giesbrecht P. (1987) Trimethoprim-induced structural alterations in *Staphylococcus aureus* and the recovery of bacteria in drug-free medium. J. Antimicrob. Chemother. 19, 147-159.

O'Neill M.A.A., Beezer A.E., Labetoulle C., Nicolaides L., Mitchell J.C., Orchard J.A., Connor J.A., Kemp R.B., Olomolaiye D. (2003) The base catalysed hydrolysis of methyl paraben: a test reaction for flow microcalorimeters used for determination of both kinetic and thermodynamic parameters. Thermochim. Acta 399, 63–71.

Ogren W. (1990) What in the world is propolis used for? Am. Bee J. 130, 239-240.

Ota C., Unterkircher C., Fantinato V., Shimizu M. T. (2001) Antifungal activity of propolis on different species of *Candida*. Mycoses 44, 375-378.

Peng Y. S., Fang Y., Xu S., Ge L., Nasr M. E. (1987) Response of foster Asian honey bee (*Apis cerana* Fabr.) colonies to the brood of European honey bee (*Apis mellifera* L.) infested with parasitic mite *Varroa jacobsoni* Oud. J. Invert. Pathol. 49, 259-264.

Pepeljnjak S., Jalsenjak I., Maysinger D. (1982) Growth inhibition of *Bacillus subtilis* and composition of various propolis extracts. Pharmazie 37, 439-440.

Pereira A.S., Bicalho B., Neto F.R.A. (2003) Comparison of propolis from *Apis mellifera* and *Tetragonisca angustula*. Apidologie 34, 291-298.

Petri G., Lembercovics E., Foldvari M. (1986) Examination of differences between propolis (bee glue) produced from different floral environments. In: Flavours and Fragrances: A World Perspective, Lawrence B.M., Mookhedjee B.D., Willis B.J. (eds.), Elsevier, Amsterdam, pp. 439–446.

Popravko S. A (1978) Chemical composition of propolis, its origin and standardization. In: A remarkable hive product: propolis, Apimondia Publ. House, Bucharest, pp. 15-18.

Popravko S.A. and Sokolov M.V. (1980) Plant sources of propolis, Pchelovodstvo 2, 28-29 (in Russian).

Prokopovich N.N. (1957) Propolis a new anaesthetic. Vrach. Delo 10, 1077-1080. (in Russian).

Prokopovich N.N., Flis Z.A., Frankovskaya Z.I., and Kope'eva E.P. (1956) An anaesthetizing substance for use in stomatology. Vrach. Delo 1, 41-44. (in Russian).

Rand J.D., Danby S.G., Greenway D.L., England R.R. (2002) Increased expression of the multi-drug efflux genes acrAB occurs during slow growth of *Escherichia coli*. FEMS Microbiol. Lett., 207, 91-95.

Rehm S. M. and Ritter W. (1989) Sequence of sexes in the offspring of *Varroa jacobsoni* and the resulting consequences for the calculation of the developmental period, Apidologie 20, 339-343.

Renobales M.D., Nelson D.R., Blomquist G.J. (1991) Cuticular lipids. In: Physiology of the insect epidermis, Binnington K. and Retnakaran A. (eds.), CSIRO, Melbourne. pp. 240-251

Rios J.L., Recio M.C., Villar A. (1988) Screening methods for natural products with antimicrobial activity: A review of the literature. J. Ethnopharm. 23, 127-140.

Ritter W. and Roth H. (1988) Experiments with mites resistance to varroacidal substances in the laboratory. In: European research on varroatosis control, Cavalloro R. (ed.), Proc. Meet. EC Experts' Group, Bad Homburg, October 1986, Balkema, Rotterdam, pp 157-160.

Rösch G.A. (1927) Beobachtung an Kitthartz sammelnden Bienen (*Apis mellifica* L.). Biolog. Zentralb. 47, 113-121.

Rosenkranz P. (1985) Temperaturpräferenz von *Varroa jacobsoni* und Verteilung des Parasiten in Brutnest von *Apis mellifera*. Apidologie 16, 213-214.

Rosenkranz P. (1987) Thermobehandlung Verdeckelter Arbeiterinnen-Brutwaben als Möglichkeit der Varroatose-Kontrolle. Apidologie 18, 385-387.

Russell A.D. (2002) Antibiotic and biocide resistance in bacteria: Introduction. J. appl. Microbiol. 92, 1S-3S.

Russell A.D. and Chopra I. (1996) Understanding antibacterial action and resistance. 2nd edn. Chichester: Ellis Horwood.

Sammataro D., Gerson U., Needham G. (2000) Parasitic mites of honey bees: life history, implications and impact, Annu. Rev. Entomol. 45, 519-548.

Santos F. A., Bastos E. M. A., Uzeda M., Carvalho M. A. R., Farias L. M, Moreira E. S. A., Braga F. C. (2002) Antibacterial activity of Brazilian propolis and fractions against oral anaerobic bacteria. J. Ethnopharm. 80, 1-7.

Sawaya A.C.H.F., Palma A.M., Caetano F.M., Marcucci M.C., da Silva Cunha I.B., Araujo C.E.P., Shimizu M.T. (2002) Comparative study of in vitro methods used to analyse the activity of propolis extracts with different compositions against species of Candida. Let. Appl. Microbiol. 35, 203-207.

Scheller S., Szaflarski J., Tustanowski J., Nolewajka E., Stojko A. (1977) Biological properties and clinical applications of propolis I. Arzneim-Forsch. Drug Res. 27, 889-890.

Schkurat B.T., and Poprawko C.A. (1980) Effect of propolis against *Varroa*, Pcelovodstvo 1, 19. (in Russian).

Schmid-Hempel P., Winston M. L., Ydenberg R. C. (1993) Foraging of individual workers in relation to colony state in the social Hymenoptera. The Can. Entom. 125, 129-160.

Schmolz E. and Lamprecht I. (2000) Calorimetric investigation of activity states and development of holometabolous insects. Thermochim. Acta 349 61-68.

Schmolz E. and Schulz O. (1995) Calorimetric investigation on thermoregulation and growth of wax moth larvae *Galleria mellonella*. Thermochim. Acta 251, 241-245.

Schmolz E., Drutschmann S., Schricker B., Lamprecht I. (1999) Calorimetric measurements of energy contents and heat production rates during development of the wax moth *Galleria mellonella*. Thermochim. Acta 337, 83-88.

Schneider P. and Drescher W. (1987) Einfluss der Parasitierung durch die Milbe *Varroa jacobsoni* Oud auf das Schlupfgewicht, die Gewichtsentwicklung, die Entwicklung der Hypopharynxdrüsen und die Lebensdauer von *Apis mellifera* L. Apidologie 18, 101-110.

Schneider P. and Drescher W. (1988) Die Folgen eines Unterschiedlich hohen *Varroa*-befalls während der Pupenentwicklung auf die erwaschene Biene. Allg. D. Imkerzeit. 22, 16–18, 54–56.

Schulz A. E. (1984) Reproduktion und Populationsentwicklung der Parasitischen Milbe *Varroa jacobsoni* Oud. in Abhängigkeit vom Brutzyklus ihres Wirtes, *Apis mellifera* L. Apidologie 15, 401-420.

Sehnal F. (1966) Kritisches Studium der Bionomie and Biometrie der in verschiedenen Lebensbedingungen gezüchteten Wachsmotte, *Galleria mellonella*. Zeitsch. Wissensch. Zool. 174, 53-83.

Semple R.L., Hicks P.A., Lozare J.V., Castermans A. (1992) Towards integrated commodity and pest management in grain storage: a Training Manual for application in humid tropical storage systems. Proceedings and selected papers from the Regional Training Course on Integrated Pest Management Strategies in Grain Storage Systems, conducted by the National Post Harvest Institute for Research and Extension (NAPHIRE), Department of Agriculture, June 6-18, 1988, Philippines. A REGNET (RAS/86/189) Publication in Collaboration with NAPHIRE. pp. 526.

Sforcin J. M., Fernandes Jr A. C., Lopes A. M., Bankova V., Funari S. R. C. (2000) Seasonal effect on Brazilian propolis antibacterial activity. J. Ethnopharm. 73, 243-249.

Shimanuki H (1981) Controlling the greater wax moth. USDA publication.

Shimanuki H., Calderone N. W., Knox D.A. (1994) Parasitic mite syndrome: The symptoms. Am. Bee J. 134, 117-119.

Shub Z.A., Kagramanova K.A., Voropaeva S.D., Kivma G.I. (1981) Effect of propolis on *Staphylococcus aureus* strains resistant to antibiotics. Antibiotiki 26, 268-271.

Simuth J. (1986) Inhibition of bacterial DNA-dependant RNA polymerases and restriction endonuclease by UV-absorbing components from propolis. Pharmazie 41, 131-32

Smirnov A.M. (1978) Research results obtained in USSR concerning disease. Etiology, pathogenesis, epizootiology, diagnosis and control of *Varroa*. Apiacta 13, 149-162.

Snodgrass R.E. (1935) Principles of Insect Morphology: Chapter III the body wall and its derivatives. Cornell university press.

Spiridonov N.A., Arkhipov V.V., Narimanov A.A., Shabalina S.A., Zverkova L.A., Shvirst E.M., Kondrashova M.N. (1992) Effect of *Galleria mellonella* larvae preparation and honey bee products on cell cultures. Comp. Biochem. Physiol.: C - Comp. Pharmacol. Toxicol. 102, 205-208.

Spreafico M., Eördegh F.R., Bernardinelli I., Colombo M. (2001) First detection of strains of *Varroa destructor* resistant to coumaphos, results of laboratory test and field trials. Apidologie 32, 49-55.

Starzyk J., Scheller S., Szaarski J., Moskwa M., Stojko A. (1977) Biological properties and clinical application of propolis. II. Studies on the antiprotozoan activity of ethanol extract of propolis. Arzneim-Forsch. Drug Res. 27, 1198-1199.

Steiner J., das Garças Pompolo S., Takahashi C.S., Gonçalves L.S. (1982) Cytogenetics of the acarid *Varroa jacobsoni*. Rev. Bras. Genet. 5, 841-844.

Steiner J., Diehl P.A., Vlimant M. (1995) Vitellogenesis in *Varroa jacobsoni*, a parasite of honey bees. Exp. App. Acar. 19, 411-422.

Steiner J., Dittmann F., Rosenkranz P., Engels W. (1994) The first gonocycle of the parasitic mite (*Varroa jacobsoni*) in relation to pre-imaginal development of its host, the honey bee (*Apis mellifera carnica*). Invert. Reprod. Devlop. 25, 175-183.

Strehl E., Volpert R., Elstner E.F. (1994) Biochemical activities of propolis extracts: III Inhibition of dihydrofolate reductase. Z. Naturforsch. 49c, 39-43.

Strong, R. G. and Dickman J. (1973) Comparative effectiveness of fifteen insect growth regulators against several pests of stored products. J. Econ. Entomol. 66, 1167-1173.

Stürz B. and Wallner K. (1997) Rückstandsuntersuchungen in Bienenprodukten. Allg. Dtsch. Imkerztg. 31, XIV-XV.

Tabor K. L., and Ambrose J. T. (2001) The use of heat treatment for control of the honey bee mites *Varroa destructor.* Am. Bee J. 141, 733-736.

Takaisi-Kikuni N.B., and Schilcher H. (1994) Electron microscopic and microcalorimetric investigations of the possible mechanism of the antibacterial actions of a defined propolis provenance. Planta Med. 60, 222-227.

Tamas M., Marinescu I., Ionescu F. (1979) Flavonoideled in muguri de plop. Stud. Cercet. Biochim. 22, 207-213.

Tegos G., Stermitz F.R., Lomovskaya O., Lewis K. (2002) Multidrug Pump Inhibitors Uncover Remarkable Activity of Plant Antimicrobials. Antimicrob. Agents Chemother. 46, 3133-3141.

Theisen M.O., Miller G.C., Cripps C., Renobales M.D., Blomquist G.J. (1991) Correlation of carbaryl uptake with hydrocarbon transport to the cuticular surface during development in the cabbage looper, *Trichoplusia ni. Pestic. Biochem. Physiol.* 40, 111-116.

Tomas-Barberan F.A., Garcia-Viguera C., Vit-Olivier P., Ferreres F., Tomas-Lorente F. (1993) Phytochemical evidence for the botanical origin of tropical propolis from Venezuela. Phytochem. 34, 191–196.

Trouiller J., Arnold G., Le Conte Y., Masson C. (1991) Temporal pheromonal and kairomonal secretions in brood of honeybees. Naturwissenschaften 78, 368-370.

Trubin A.V., Chernov K.S., Kuchin L. A., Borzenko I.E., Yalina A.G. (1987) European foulbrood: transmission and sensitivity of the causal agents to antibiotics. Veterinariya 8, 46-47, (in Russian).

Van Ketel W. G. and Bruynzeel D. P. (1992) Occupational dermatitis in an accordion repairer. Cont. dermat. 27, 186.

Wadsö I. (2002) Isothermal microcalorimetry in applied biology. Thermochim. Acta 394, 305-311.

Walker C.B. (1996) The acquisition of antibiotic resistance in the periodontal microflora. Periodontology 2000 10, 79-88.

Wallner K. (1991) Das Verhalten von Paradichlorbenzol in Wachs and Honig. Allg. Deut.. Imk. Zeit. 9, 29-31.

Wallner K. (1995) Nebeneffekte bei Bekämpfung der Varroamilbe. Die Rückstandssituation in einigen Bienenprodukten. Bienenvater 116 (4), 172-177.

Wallner K. (1999) Varroacides and their residues in bee products. Apidologie 30, 235-248.

Weinberg K. P., and Madel G. (1985) The influence of the mite *Varroa jacobsoni* Oud. on the protein concentration and the hemolymph volume of the brood of worker bees and drones of the honeybee *Apis mellifera* L. Apidologie 16, 421-436.

Wiegers F. P. (1986) Transmission of acute paralysis virus by the honeybee parasite *Varroa jacobsoni* Oud. In: Fundamental and applied aspects of invertebrate pathology, Samson R. A., Vlak J. M., Peters D. (eds.) pp. 152, Wageningen.

Williams C. M (1967) Third Generation Pesticides. Sci. Am. 217, 13-17.

Williams D. and Amos T. G. (1974) Some effects of synthetic juvenile hormones and hormone analogues on *Tribolium castaneum* (Herbs). Aust. J. Zool. 22, 147-53.

Winston M.L. and Fergusson L.A. (1985) The effect of worker loss on temporal caste structure in colonies of the honeybee. Can. J. Zool. 63, 777–780.

Woisky R. and Salatino A. (1998) Analysis of propolis: some parameters and procedures for chemical quality control. J. Apicult. Res. 37, 99-105.

Wollenweber E. and Buchmann S.L. (1997) Feral honey bees in the Sonoran Desert: propolis sources other than poplar (*Populus* spp.). Z. Naturforsch. 52c, 530–535.

Yakobson B., Navarro S., Donahaye E.J., Azrieli A., Slaveski Y., Ephrati H. (1997) Control of beeswax moths using carbon dioxide in flexible plastic and metal structures. In: Proc. int. conf. controlled atmosphere and fumigation in grain storages, 21- 26 April 1996, Nicosia, Cyprus, 169-174.

Zgurskaya H. I. and Nikaido H. (1999). Bypassing the periplasm: reconstitution of the AcrAB multidrug efflux pump of *Escherichia coli*. Proc. Natl. Acad. Sci. USA 96,7190–7195.

12. Appendix

12.1 Appendix to Chapter 7

Table A1 Antimicrobial activities of different concentrations of EEP (% w/v) from different geographic origins against bacteria and a yeast displayed by the mean diameter of the inhibition zone (mm). n = 3, l.g. stands for lawn of bacterial growth due to stimulation by the effect known as hormesis.

E. coli

Propolis type	Inhibition Diameter		
	0.1%	1.0%	10.0 %
WEP	0.00	0.00	0.00
I1	0.00	3.40	5.50
E1	0.00	0.00	0.00
C1	0.00	0.00	0.00
K1	1.97	13.53	16.57
G1	0.00	3.40	6.83
RUS1	0.00	0.00	24.00
P1	3.07	8.57	11.67
P2	1.73	2.80	5.80
P3	2.33	7.10	10.07
P4	2.37	6.10	10.80
SA1	1.63	2.47	5.67
SA3	0.00	0.00	6.87
SA5	0.00	5.77	5.80
SA6	0.00	0.00	4.40
SA8	0.00	2.13	4.70
SA11	0.00	2.50	6.77

B. brevis

Propolis typ	Inhibition Diameter			
	0.01%	0.1%	1.0%	10.0 %
WEP	0.00	0.00	3.61	9.30
I1	1.63	9.12	13.87	18.07
E1	5.77	8.30	15.57	23.40
C1	0.00	9.83	12.37	22.47
K1	4.43	16.80	29.43	31.18
G1	3.30	9.12	13.87	20.40
RUS1	4.43	11.13	17.80	18.30
P1	4.30	12.57	18.80	24.13
P2	3.60	10.00	16.83	23.50
P3	0.00	11.30	17.23	21.50
P4	0.00	11.60	19.13	25.10
SA1	9.00	15.03	20.47	22.90
SA3	3.07	13.80	20.40	24.60
SA5	8.90	14.57	18.33	24.63
SA6	9.50	14.23	21.20	24.40
SA8	9.80	13.50	23.33	28.33
SA11	5.67	10.83	20.40	26.03

M. lutes

Propolis typ	Inhibition Diameter			
	0.01%	0.1%	1.0%	10.0 %
WEP	0.00	0.00	0.00	5.67
I1	0.00	3.43	7.60	11.33
E1	0.00	0.00	0.00	8.87
C1	0.00	0.00	6.20	10.27
K1	0.00	3.43	7.00	11.00
G1	0.00	6.83	15.90	16.77
RUS1	0.00	9.37	7.97	11.34
P1	0.00	4.87	11.23	12.70
P2	0.00	0.00	9.03	10.13
P3	0.00	4.63	10.87	11.80
P4	0.00	5.17	10.20	13.40
SA1	0.00	10.23	17.13	17.00
SA3	0.00	6.33	11.53	17.33
SA5	4.50	11.57	15.43	18.87
SA6	4.50	10.73	13.43	19.47
SA8	0.00	7.73	14.70	18.80
SA11	0.00	3.20	11.43	18.67

B. megaterium

Propolis typ	Inhibition Diameter			
	0.01%	0.1%	1.0%	10.0 %
WEP	0.00	0.00	0.00	3.60
I1	2.07	4.97	8.07	10.77
E1	0.00	5.27	10.53	12.93
C1	0.00	4.77	12.67	12.07
K1	0.61	6.40	15.40	16.30
G1	2.27	4.63	8.07	10.43
RUS1	2.50	8.47	8.57	11.40
P1	2.03	6.10	10.20	12.97
P2	1.63	5.87	10.67	15.43
P3	2.37	7.43	11.30	12.07
P4	2.23	7.17	16.17	15.50
SA1	3.23	11.40	21.83	18.33
SA3	0.00	6.50	11.33	16.20
SA5	5.17	10.87	14.83	19.00
SA6	3.50	11.03	13.77	17.70
SA8	3.17	11.43	14.47	20.07
SA11	0.00	7.40	17.17	22.00

Table A1 contd.

B. subtilis					S. cerevisae				
Propolis	Inhhibition Diameter				Propolis	Inhhibition Diameter			
typ	0.01%	0.1%	1.0%	10.0 %	typ	0.01%	0.1%	1.0%	10.0 %
WEP	0.00	0.00	0.00	7.87	WEP	0.00	0.00	0.00	5.67
I1	0.00	5.77	11.73	16.93	I1	0.00	3.00	11.87	21.37
E1	0.00	6.87	11.30	16.87	E1	0.00	0.00	5.37	6.93
C1	0.00	6.23	16.13	18.03	C1	0.00	2.23	3.87	5.60
K1	0.00	7.73	18.60	25.97	K1	0.00	16.37	33.67	50.67
G1	0.00	5.77	11.73	14.47	G1	0.00	3.00	13.57	25.37
RUS1	0.00	9.20	10.93	15.60	RUS1	0.00	0.00	6.77	15.67
P1	0.00	8.35	13.23	18.00	P1	0.00	9.60	24.53	39.60
P2	0.00	6.20	9.47	15.13	P2	0.00	0.00	13.90	36.00
P3	0.00	8.80	11.83	16.50	P3	0.00	9.43	17.83	52.67
P4	0.00	7.67	14.80	15.67	P4	0.00	4.50	22.77	44.13
SA1	4.60	12.30	18.07	28.67	SA1	0.00	3.60	6.47	23.97
SA3	0.00	7.43	13.30	16.70	SA3	0.00	0.00	3.20	7.20
SA5	4.90	11.03	17.20	21.40	SA5	0.00	8.20	21.77	23.90
SA6	5.00	11.10	15.33	21.67	SA6	0.00	7.07	13.00	23.70
SA8	3.60	9.53	18.67	22.00	SA8	1.50	6.33	12.47	25.53
SA11	0.00	6.03	11.50	18.63	SA11	0.00	0.00	13.43	22.13

P. syringae

Propolis	Inhhibition Diameter				
typ	0.001%	0.01%	0.1%	1.0%	10.0 %
WEP	0.00	0.00	0.00	0.00	7.87
I1	0.00	0.00	5.00	9.33	13.00
E1	l.g	l.g	4.1 + l.g.	11.60	13.33
C1	0.00	l.g	3.7 + l.g.	9.70	12.00
K1	l.g	l.g	4.6 + l.g.	21.60	22.33
G1	l.g	l.g	l.g	10.67	15.00
RUS1	0.00	0.00	8.80	12.67	16.33
P1	l.g	l.g	5.8 + l.g.	15.57	19.00
P2	0.00	l.g	l.g	10.17	13.67
P3	0.00	l.g	3.3 + l.g.	12.20	15.00
P4	0.00	l.g	4.10	15.53	20.00
SA1	l.g	4.8 + l.g.	17.80	22.60	27.67
SA3	l.g	l.g	4.3 + l.g.	10.50	16.00
SA5	l.g	7.0 + l.g.	17.50	24.28	25.67
SA6	0.00	0.00	0.00	0.00	0.00
SA8	l.g	3.3 + l.g.	10.43	20.60	27.33
SA11	0.00	l.g	6.4 + l.g.	17.88	27.33

Table A2 Antifungal activities of different concentrations of EEP (% w/v) from different geographic origins as displayed by the mean diameter of the inhibition zone (mm). n = 3.

A. niger

Propolis type	Inhibition Diameter				
	1%	2%	4%	8%	10%
WEP	0.00	0.00	0.00	0.00	0.00
I1	3.50	7.67	6.67	10.33	12.67
E1	0.00	0.00	5.33	8.67	8.67
C1	4.00	7.00	7.00	7.67	8.00
K1	8.00	15.33	13.00	13.00	13.00
G1	0.00	7.67	6.67	10.33	11.67
RUS1	3.50	6.00	7.33	9.00	9.33
P1	2.50	7.00	9.33	9.33	12.33
P2	3.00	8.00	12.00	11.33	14.67
P3	3.30	8.33	9.00	10.67	11.00
P4	4.50	12.67	14.00	13.00	12.00
SA1	2.90	6.33	6.67	8.33	9.33
SA3	1.50	4.33	5.00	6.00	7.67
SA5	4.00	7.00	11.33	11.00	11.00
SA6	2.50	5.67	8.00	7.00	8.67
SA8	3.50	6.33	6.33	6.00	6.67
SA11	2.90	5.67	6.00	6.00	5.67

P. chrysogenum

Propolis type	Inhibition Diameter				
	1%	2%	4%	8%	10%
WEP	0.00	0.00	0.00	0.00	0.00
I1	3.00	7.67	10.67	12.33	11.00
E1	2.50	8.00	6.00	7.00	7.67
C1	0.00	5.00	5.00	6.33	7.00
K1	7.50	12.00	14.33	12.67	16.33
G1	0.00	7.67	10.67	12.33	11.00
RUS1	0.00	3.00	6.00	7.33	8.33
P1	3.30	8.33	11.00	12.00	13.67
P2	7.00	11.67	12.00	13.00	14.00
P3	6.00	11.00	13.00	15.33	18.67
P4	5.70	10.67	13.67	12.00	12.33
SA1	2.00	5.00	8.00	9.00	10.67
SA3	3.10	8.00	8.00	8.33	8.67
SA5	2.60	6.67	7.00	8.33	12.00
SA6	1.50	6.00	6.00	6.00	10.00
SA8	3.90	7.33	7.67	6.67	7.00
SA11	0.00	3.00	3.67	5.00	4.67

Table A2 Contd.

T. viridae

Propolis type	Inhibition Diameter				
	1%	2%	4%	8%	10%
WEP	0.00	0.00	0.00	0.00	0.00
I1	3.50	7.67	8.67	8.00	10.00
E1	0.00	0.00	0.00	0.00	0.00
C1	0.00	0.00	0.00	0.00	0.00
K1	5.10	12.67	17.00	14.33	14.33
G1	2.50	7.67	8.67	8.00	16.00
RUS1	0.00	4.67	6.00	8.33	8.67
P1	3.00	9.00	16.67	17.67	19.33
P2	4.00	8.00	10.00	12.67	12.00
P3	6.00	10.67	11.00	14.00	13.00
P4	7.50	11.33	11.00	9.67	12.67
SA1	0.00	0.00	0.00	0.00	0.00
SA3	0.00	0.00	0.00	0.00	0.00
SA5	0.00	0.00	0.00	0.00	3.33
SA6	0.00	0.00	0.00	0.00	0.00
SA8	0.00	3.33	6.00	5.33	6.00
SA11	0.00	3.33	4.00	5.33	6.67

Table A3 Antimicrobial activities of the volatile components of propolis (PV)
Activities of propolis from different geographic origins demonstrated by the mean inhibition diameter (mm) of a 10% (w/v) PV. n = 3.

	E1	C1	G1	P1	P2	SA1	SA3	SA5	I1
B. brevis	20.15	19.53	18.55	21.11	19.57	18.78	21.82	20.19	19.42
B. megaterium	9.22	10.86	9.51	11.23	13.56	16.43	14.32	15.87	11.32
B. subtilis	14.33	15.32	12.76	15.01	15.11	25.86	14.76	18.55	16.45
M. luteus	6.56	8.59	16.22	11.60	9.15	16.45	15.55	16.67	13.24
E. coli	n.d.	n.d.	5.93	10.65	5.23	4.44	5.81	5.12	8.67
P. syringae	10.45	11.52	13.25	16.77	11.25	24.45	14.39	22.35	13.45
S. cerevisae	5.76	5.12	20.57	35.42	31.34	21.95	7.11	20.58	6.48
A. niger	11.52	10.25	10.34	9.55	10.17	10.23	7.67	5.67	11.25
P. chrysogenum	15.11	9.53	11.54	15.64	10.11	9.95	9.12	5.58	9.35
T. viridae	12.54	13.35	17.67	11.56	12.35	2.85	n.d.	5.00	13.86

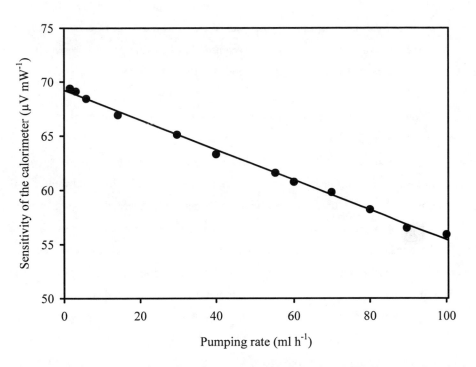

Fig. A1 Electrical calibration of the flow calorimeter (10700-1, LKB Bromma, Sweden) whilst circulating a phosphate buffer of pH 7.0 at different pumping rates and at a temperature of 30 °C.

12.2. Acknowledgments

Several people have contributed to the success of this thesis in one way or another.

I am grateful to **Prof. Dr. Burkhard Schricker** to welcome me in his research group and provide me with the necessary materials and space required through out my thesis research. His continuous encouragement and motivation during the write-up of this thesis had an immense input.

The contributions of **Prof. Dr. Ingolf Lamprecht** to the success of my thesis work from the very beginning to the end is invaluable. His contribution, first of all, made it possible for me to come to Germany for the first time in the year 1995 to do the experimental part of my M.Sc. thesis in the fields of biocalorimetry and propolis research. The scientific knowledge and experience I obtained during that time and the social atmosphere in his research group influenced my interest substantially and motivated me to continue my scientific career in the aforementioned fascinating fields. His contributions in problem identification, openness for discussion at any time, his invaluable comments and suggestions led to the success of my thesis research. His willingness to proof read manuscripts, important suggestions and comments played vital role in the publication of several articles during my thesis work. His help was not only scientifically, he also helped me in integrating in the German culture; by frequently inviting me in his very guest friendly family showing that he was more than a supervisor: a mentor and a friend. I would also like to take this opportunity to express my gratefulness to his wife Natasha.

Dr. Erik Schmolz contributed to the success of my thesis substantially. His readiness to discuss and solve problems, and his important suggestions and comments were invaluable through out my work. His help was not limited only to the laboratory work and solving scientific problems, his cheerful nature also had a great input in easing a stressed mind. His effort in making the working atmosphere suitable by supplying the necessary computer, software, and other facilities played crucial roles. The concerts and parties in his cellar with his Music band "the Gogomaniacs", Erik being the main singer and guitarist, played remarkable roles in freeing the mind from the day to day laboratory routine.

I thank **Dr. Benedikt Polaczek** for his relentless help in supplying me Varroa infested brood combs and in the collection of mites. Though the collection of mites from infested brood was not a pleasant job to do, he was never reluctant to help. I am also very grateful to him for supplying me with the propolis samples from Germany and Poland. Apart from that his non-stop supply of honey to make my daily coffee sweet had an input in my work.

My heart-felt gratitude goes to **Prof. Dr. Randall Hepburn, Dr. Eyualem Abebe, and Dr. Diana Alexandra Torres Sanchez** for supplying me with propolis samples from South Africa, Ethiopia, and Colombia, respectively.

Microbiological experiments would have been very difficult without the friendly laboratory space provided to me by **Dr. Lars Podsiadlowski**. In addition to that his readiness to help played vital roles in the success of the microbiological work.

I am grateful to **Prof. Dr. Klaus Hausmann** for allowing me to use the electron microscope of his research group to take some important pictures of Varroa mites.

The technical assistance provided to me by **Mrs. Gudrun Welge, Mrs. Beate Bach,** and **Mrs. Christine Bergmann** was enormous. They were always cheerful and ready to help.

I am grateful to **Thorsten Scholz** for his help in solving the computer problems that arose frequently due to the shared use of a PC at the beginning of my PhD work.

I thank **Dr. Klaus Wallner** for his help in analysing the propolis samples for acaricidal residues.

My genuine thanks go to my friends **Dr. Inge Steinmetz, Dr. Robbie Aitken,** and **Silke Beckedorf** for proof reading the final version of the thesis and for the useful comments and suggestions they gave me to make the thesis clearly readable and easily understandable. Especially Inge and Silke played crucial role in the proper construction and finalization of the German version of the general discussion.

My special thanks go to **Patricia Codyre** for her patience and endurance in taking care of our son Lucas in the metropolitan N.Y. city whilst I was concentrating on my thesis research. I also owe her thanks for designing and constructing my personal URL, currently at http://userpage.fu-berlin.de/~assegid/.

I thank **Dr. Monika Normant** and **Claudia Contzen** for their nice laboratory and office company and the pleasant social atmosphere we had.

I am very grateful to the German Academic Exchange Service **(DAAD)** for the financial support, without which my stay in Germany would have been difficult.

I thank **Roisin Codyre** and **Thomas Codyre** for their warm welcome and extreme guest friendliness during my frequent visits in Ireland.

Last but not least I thank my **mom Yeshi Gebrie,** my **brothers,** and **sisters** for their never-ending moral support.

12.3 Personal Data

Assegid Garedew, born on 01.05. 1969, in Northern Showa, Ethiopia.

Education and qualifications

Sept.1983 – Mar.1987 **High school** attended at Hailemariam Mammo Comprehensive Secondary School, Debrebirhan and Completed with distinction

Sept.1987 - Dec.1991 Undergraduate study at Addis Ababa University, Faculty of Natural Science. Earned a **B.Sc. Degree in Biology** with distinction and awarded the University gold medal upon graduation being the outstanding student of the year.

Sept.1993 - July 1996 Graduate Study at the School of graduate studies, Addis Ababa University Earned a **M.Sc. Degree in Biology** with distinction and Awarded the "Ethiopian Scientific Society (ESS) - Student Academic Excellence Award" of the year 1995.

Apr. – Oct. 1995 Conducted research in the Institute for Biophysics, Free University of Berlin, as part of my M.Sc. thesis.

Apr. 2000 – Nov.2003 **Ph.D study** at the Faculty of Biology, Chemistry, and Pharmacy, Institute of Zoology, Free University of Berlin, Germany.

Professional experience

March 1992 - June 1993   Graduate Assistant at the Bahir Dar Teachers College, involved in teaching different Biology courses for diploma students.

July 1993 - Sept. 1996   Assistant Lecturer at the Bahir Dar Teachers College, involved in teaching different Biology courses for diploma students.

Oct. 1996 – September 1999   Lecturer at the Bahir Dar University, involved in teaching different Biology courses for undergraduate students and conducting research.

12.4 List of own publications

Assegid Garedew, Ingolf Lamprecht, Erik Schmolz, Burkhard Schricker (2002) The varroacidal action of propolis: a laboratory assay. Apidologie 33, 41-50.

Assegid Garedew, Erik Schmolz, Burkhard Schricker, Ingolf Lamprecht (2002) Microcalorimetric investigation of the action of propolis on *Varroa destructor* mites. Thermochimica Acta 382, 211-220.

Assegid Garedew, Erik Schmolz, Ingolf Lamprecht (2003) Microcalorimetric and respirometric investigation of the effect of temperature on the antivarroa action of Propolis. Thermochimica Acta 399, 171–180.

Assegid Garedew, Erik Schmolz, Burkhard Schricker, Benedikt Polaczek, Ingolf Lamprecht (2002) Energy metabolism of *Varroa destructor* mites and its implication on host vigour. J. Apicult. Sci. 46, 73-83.

Assegid Garedew, Erik Schmolz, Burkhard Schricker, Ingolf Lamprecht –The energy and nutritional demand of the parasitic life of the mite *Varroa destructor*. (in press: Apidologie).

Assegid Garedew, Ingolf Lamprecht, Erik Schmolz, Burkhard Schricker (2002) Microcalorimetric toxicity investigation of propolis on *Tenebrio molitor* L. (Coleoptera: Tenebrionidae). Thermochimica Acta, 394, 239-245.

Assegid Garedew, Erik Schmolz, Ingolf Lamprecht (2003) The antimicrobial activity of honey of the stingless bee *Trigona spp* and the implication of its use as a panacea. J. Apicult. Sci. 47, 37-49.

Assegid Garedew and I. Lamprecht (1997) Microcalorimetric investigations on the influence of propolis on the bacterium *Micrococcus luteus*. Thermochimica Acta 290, 155-166.

Assegid Garedew, Erik Schmolz, Ingolf Lamprecht, The effect of propolis on the metabolic rate and metamorphosis of the greater wax moth *Galleria mellonella* (accepted: Thermochimica Acta).

Assegid Garedew, Erik Schmolz, Ingolf Lamprecht, Microcalorimetric investigation on the antimicrobial activity of honey of the stingless bee *Trigona spp.* and comparison of some parameters with those obtained with standard methods (in press: Thermochimica Acta).

Alexandra Torres, Assegid Garedew, Erik Schmolz, Ingolf Lamprecht, Calorimetric investigation of the antimicrobial action and insight into the chemical properties of "angelita" honey – a product of the stingless bee *Tetragonisca angustula* from Colombia. (in press: Thermochimica Acta).

Claudia Contzen, Assegid Garedew, Ingolf Lamprecht, Erik Schmolz, Calorimetric and biochemical investigations on the influence of the parasitic mite *Varroa destructor* on the development of honeybee brood. (in press: Thermochimica Acta).

Assegid Garedew, Erik Schmolz, Ingolf Lamprecht, Microcalorimetric and microbiological investigations on the antimicrobial actions of Kazakh mumiyo. (submitted: J. Ethnopharm.).

Assegid Garedew, Michael Feist, Erik Schmolz, Ingolf Lamprecht, Thermal analysis of mumiyo, the legendary folk remedy from the Himalaya region. (Accepted: Thermochimica Acta).

Assegid Garedew, Erik Schmolz, Ingolf Lamprecht, Microbiological and calorimetric studies on the antimicrobial actions of different extracts of propolis: an *in vitro* investigation. (Submitted: Thermochimica Acta).

Assegid Garedew, Alexandra Torres, Erik Schmolz, Ingolf Lamprecht, Comparative microcalorimetric and antimicrobial investigations of different stingless bee honeys and elucidation of the unique chemical nature responsible for their biological activity. (Submitted: Thermochimica Acta).

Assegid Garedew, Erik Schmolz, Ingolf Lamprecht, Comparative investigations of the antivarroa actions of propolis from different geographic origins. (in preparation).

Assegid Garedew, Erik Schmolz, Burkhard Schricker, Benedikt Polaczek, Ingolf Lamprecht (2002) The effect of temperature on the antivarroa action of propolis. (Abstract of conference proceeding), Apidologie 33, 478-480.

Erik Schmolz and Assegid Garedew (2002) The energetics of *Varroa destructor* on honeybee development. (Abstract of conference proceeding), Apidologie 33, 481-482.

Claudia Contzen, Assegid Garedew, Erik Schmolz, The influence of Varroa mites on energy content, hemolymph volume and protein concentration of bee pupae. (Abstract of conference proceeding), in Press: Apidologie.

Assegid Garedew, The effect of propolis on larval development and pupal metamorphosis of *Galleria mellonella*. (Abstract of conference proceeding), in Press: Apidologie.

Assegid Garedew (2003) Nicht essbar aber heilsam: in Äthiopien dient Honig von Stachellosebienen als Medizin. Deutsches Bienenjournal, October, 20-21.

12.5 Declaration (Erklärung)

Ich versichere, dass ich diese Arbeit selbstständig angefertigt, nur die angegebenen Hilfsmittel verwendet und noch nicht für Examenszwecke benutzt habe.

September 22, 2003, Berlin

Assegid Garedew